What a Performance

Key Stage 3

Peter Branson

CAMBRIDGE
UNIVERSITY PRESS

S · T · E · P

5 16

DESIGN AND TECHNOLOGY

S · T · E · P

Published by the Press Syndicate of the
University of Cambridge
The Pitt Building, Trumpington Street, Cambridge
CB2 1RP
10 Stamford Road, Oakleigh, Victoria 3166,
Australia

First published 1992

In association with Staffordshire County Council

Designed and produced by Gecko Limited,
Bicester, Oxon

Printed in Great Britain at the University Press,
Cambridge

A catalogue record for this book is available from
the British Library

ISBN 0 521 40637 4

Cover illustration by Sue Faulks

Acknowledgements

The STEP team would sincerely like to thank the
following for their help during the research and writing
of this book.

At the New Victoria Theatre, Newcastle-under-Lyme:

Peter Cheeseman (Theatre Director), Paul Jones
(Production Manager) and all those staff in the
different departments who willingly gave up their time.

At the BBC Pebble Mill Studios, Birmingham:

Richard Taylor (Engineering and Development), Paul
Howells (Graphic Design) and all the staff in the
different departments who willingly gave up their time.

Key: t = top, b = bottom, c = centre, l = left,
r = right, * = background

J. Allan Cash   8bl, 42cr
BBC   24b, 54, 55, 64t, 72c, 73cl, 88b
Bruce Coleman Ltd   81t Kim Taylor, 82t Johnny
    Johnson
Catherine Ashmore 21b
Clive Barda   21c, 38br London Performing Arts
    Library
Christopher Coggins   6bl, 7tl, 9, 11b, 12, 16, 26t,
    29, 33, 35bl, 36, 38bl, 42b, 43, 44b, 45, 47, 48,
    49, 50, 51, 58, 59, 64b, 65, 76c, 77, 78, 79tr, 79cl,
    79c, 79bc, 79br (Radios supplied by Handy Stores
    Bicester) 84, 85t, 91, 93
City of Nottingham Design Studio   88t
Donald Cooper Photostage   7tr, 7*, 25tr
Donald Cooper Photostage/Rex Features   25b
Kobal Collection   24t, 25tl, 25tc, 25*, 26tc, 26bc,
    34t 1976 National Film Trustee Co. Ltd, 38cr
    Guild Film Distribution, 39, 40, 41, 41*, 44t, 53,
    53cr Universal City Studios
Rex Features   14t, 15c, 26b Fotos International,
    62cr, 74cr, 83c
Rex Features/SIPA Press   14bl, 14br, 15t, 62t, 66b,
    83tl J. Olive
Robert Harding Picture Library   6br, 8br, 11tr, Rolf
    Richardson, 13bl, 13br, 15bl, 24c A. Evans, 25c,
    34b Lanchester Marionettes, 68tr Robert McLeod,
    68c, 74t Don Williams, 80b Chris Laurens,
    82b Robert McLeod
Rob Judges   7c
Science Photo Library   63, 66t Adam Hart-Davis, 66c
    Peter Menzel, 67t James Stevenson 67b Francoise
    Sauze, 72t Adam Hart-Davis, 79tl, 79bl Jean-Loup
    Charmet, 80* Professor Harold Edgerton, 85c Hank
    Morgan, 85b James King-Holmes
The Image Bank   42t Anthony Boccaccio, 42cl
    Michael Salas, 62cl V. Schiller, 69t Jeff Smith, 69c
    Lou Jones, 69b Michael Melford, 70 Michael Salas,
    76tr Steve Niedorf
ZEFA Ltd.   6tl, 6c, 7bl Orion Press, 7br, 8tr Robert
    Hetz, 10, 13bc Orion Press, 35br Voigt, 67*, 68tc
    Cebrian, 76tc Rauschenbach, 80t Ed Bock, 81br,
    83b

Picture Research by Jane Duff and Libby Howless

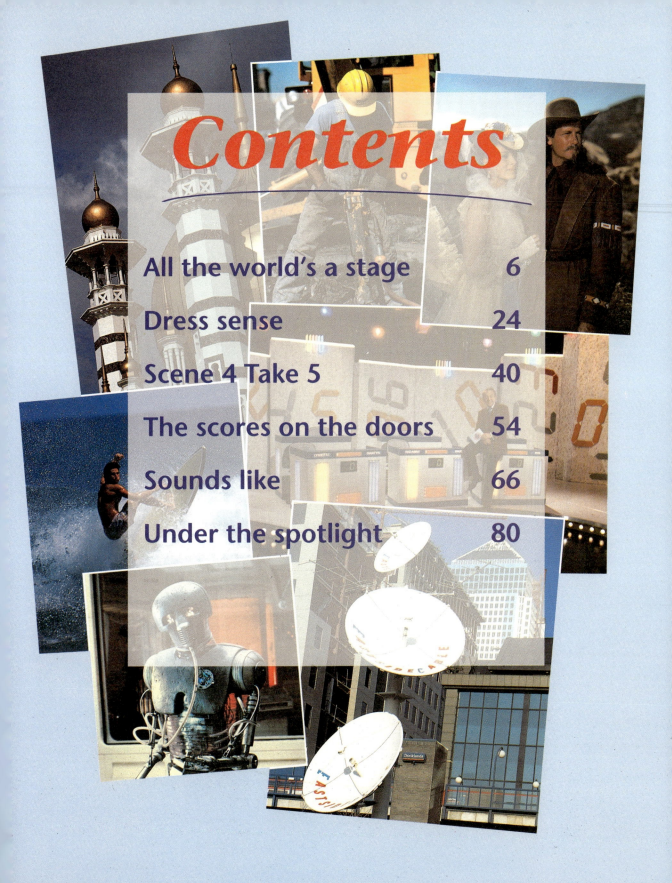

Contents

Introduction

This book takes an in-depth look at some of the activities that come together to produce a performance. Such a performance could be in the theatre, a film, on television or on radio. Most of the activities are related to what goes on 'behind the scenes'. We hope that you will begin to see how much work goes into making a television programme or putting on a radio programme. Many people are involved in all of these productions: there are lighting technicians, set designers, costume designers, script writers, production managers, carpenters, electricians, cameramen/women, wardrobe managers, film and video editors and dozens more.

Through the activities in this book, you will be involved with some of the things that they do. There will be plenty of scope for you to use your own imagination to design and make. We hope that you will get a feel for what actually goes on 'backstage'. We hope you enjoy the technological experience!

Most pages refer you to the Datafile. The names of the sheets you might need from the Datafile are shown on each page. This Datafile should be available to you within your working area. Use it to collect information when you need it for your project. Your teacher may have other information available in the form of books, magazines, computer databases, etc., for you to use.

There is a range of different activites in this section of the book. All of these activities or projects are related to some aspect of the theatre. Your teacher can give you a clearer picture of how the various parts of the section fit together. Some pages contain clearly set tasks, on other pages there are some fairly short activities, and on others there are opportunities for longer projects.

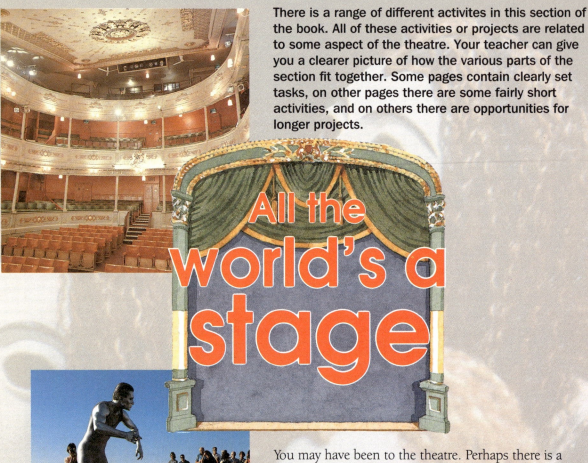

All the world's a stage

You may have been to the theatre. Perhaps there is a theatre near where you live. There are, in fact, many different types of theatre. The pictures on this page show a variety of theatres. Theatres are used for many different types of performances – plays, musicals, orchestral evenings, opera, dance, pantomime, arts festivals, variety shows, cabaret, popconcerts and comedy. You may have seen or been to a street performance or an open-air play in a park.

Perhaps your school has put on a play or a pantomime recently.

Here is a list of themes which are all linked to the theatre. You may be able to identify aspects of these themes which would be worth investigating further. There are many opportunities and needs within them.

   costumes
   seating
   advertising
   ticket distribution
   scenery and props
   lighting
   characters

You will probably be able to think of other interesting themes, related to the theatre, for yourself.

LOOKING AT
structures

Structures are very important in the theatre. They are used in building theatres and in building scenery. Structures can be used to build extensions to a stage to provide extra space.

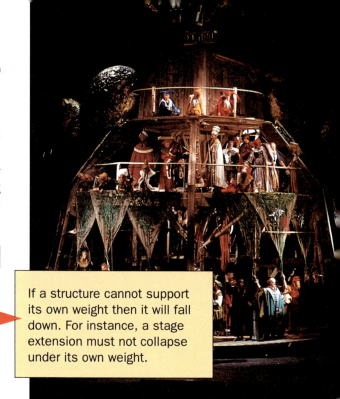

WHAT IS A STRUCTURE ?

A structure is best defined by describing what it does. A structure must be able to:

→ *support itself*

→ If a structure cannot support its own weight then it will fall down. For instance, a stage extension must not collapse under its own weight.

→ *support an object or a weight.*

→ *resist forces on it.*

The roof structure in a theatre has to support the weight of any tiles, or slates (and any extra weight from snowfall, for example). The overhead framework in a theatre must be strong enough to hold up banks of spotlights and other lighting.

Some structures are subject to forces which change. The stage in a theatre will have people and scenery constantly moving over it. The stage structure must be strong enough to cope with this. The portable seating of open-air theatres must be able to cope with the weight of people constantly moving on, off and over it.

You can see different types of structures all around you. Look at how different structures fulfil the functions listed above.

Think of other structures and decide how they are used.

Many of these complicated structures are based on simple shapes such as triangles or squares. The individual parts of any structure are called **members**.

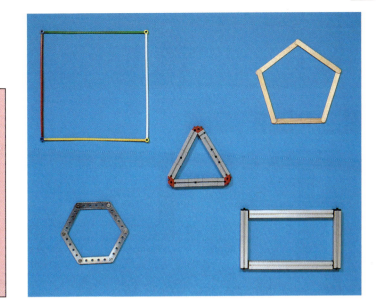

Make the structures shown on this page. They can be made with construction kits (Meccano, Fischertechnic, Polymek etc.), spaghetti, plastic strips, lolly sticks or art straws.

Apply a 'push' and a 'pull' force to each structure. A push force puts a structure, or a member, into **compression**. A pull force puts a structure, or a member, into **tension**.

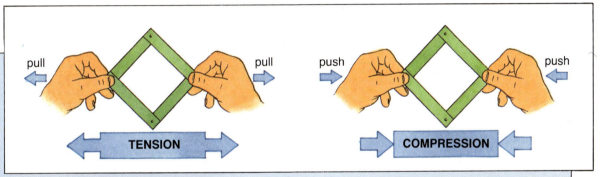

pull — pull — **TENSION**

push — push — **COMPRESSION**

Which structures are rigid or stiff? Which are flexible or able to change shape? Which structure is the strongest?

If adding extra members strengthens a structure these extra members are called **ties** or **struts**. Ties keep a structure in shape when it is under tension. Struts keep a structure in shape when it is under compression. Add extra members to some of the structures you have made. Do they become more rigid?

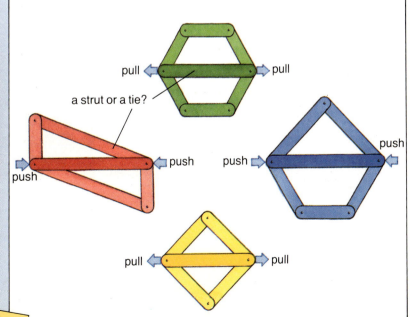

pull — pull

a strut or a tie?

push — push

push — push

pull — pull

·D A T A   F I L E·
Structures

LOOKING AT
beams & trusses

BEAMS

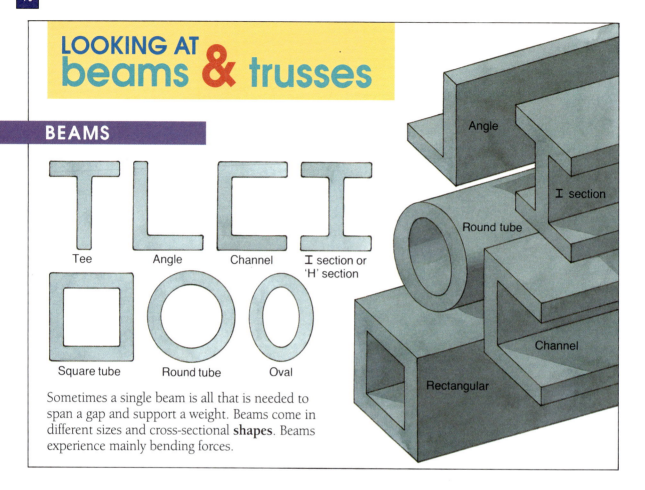

Tee

Angle

Channel

I section or 'H' section

Square tube

Round tube

Oval

Angle

I section

Round tube

Channel

Rectangular

Sometimes a single beam is all that is needed to span a gap and support a weight. Beams come in different sizes and cross-sectional **shapes**. Beams experience mainly bending forces.

In the building shown here, the roof is held up by a series of single spans across the width of the building. The weight of the roof has been calculated, a weight safety margin has been added, and a material for the beams has been chosen which is strong enough to support this total weight.

Total weight which can be supported on beams = weight of roof + weight safety margin

The upright supports in the picture have to support an even greater weight. Here, the weight of the roof beams must be added in as well. Upright beams or pillars experience compression forces.

Total weight which can be supported by uprights = weight of roof + weight of beams in roof + weight safety margin

TRUSSES

You may not want to use a single beam to span a gap. You may not have the right material to do the job or the material available may be too heavy or bulky. One way to overcome these problems is to put together several small beams or members into a larger structure. Several beams can be put together to form a truss. A truss is a structure which is longer than any of the individual beams used to make it.

(a) Lattice truss

Make some of the trusses shown this page. They can be made with construction kits (Meccano, Fischertechnic etc), spaghetti, plastic strips, lolly sticks or art straws. Investigate their strength and load-bearing properties.

(c) Warren truss

(b) Howe truss

(d) Pratt truss

(e) Bollman truss

·DATA FILE·

Structures
Materials

Here are some activities related to two very different types of theatre.

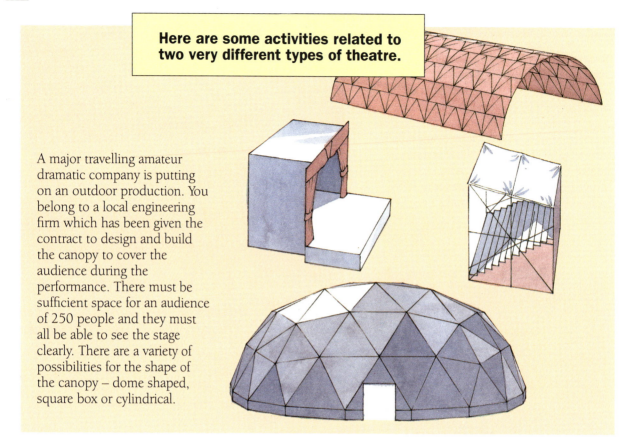

A major travelling amateur dramatic company is putting on an outdoor production. You belong to a local engineering firm which has been given the contract to design and build the canopy to cover the audience during the performance. There must be sufficient space for an audience of 250 people and they must all be able to see the stage clearly. There are a variety of possibilities for the shape of the canopy – dome shaped, square box or cylindrical.

The New Victoria Theatre in Newcastle-under-Lyme, Staffordshire, is an 'in the round' theatre. The acting area is in the middle of the auditorium. Lighting and props for a performance can be hung from a steel framework, or gantry, high up in the roof above the stage. This gantry is about 10 metres square and supports about 750 kg of lighting equipment. Can you model a suitable structure which would be light enough yet strong enough?

·D A T A   F I L E·
Structures
Fair testing
Materials

TESTING THE
strength
OF STRUCTURES

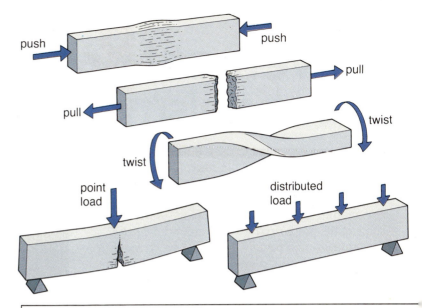

You will need to test the strength of any material you choose for your design. The structure you design will need to withstand different forces. Sometimes these may be pushing or pulling forces, sometimes they may be bending or twisting forces. You will need to carry out tests to find out what the most suitable material will be for different parts of your structure.

There are several different ways of testing structures for strength.

You may want to:
● load your structure at a particular point,
● load your structure over its length (a distributed load),
● carry out a non-destructive test,
● carry out a destructive test.

You will also need to think about how the different parts of your structure will be joined together. Here are some ways of joining different materials to create structures.

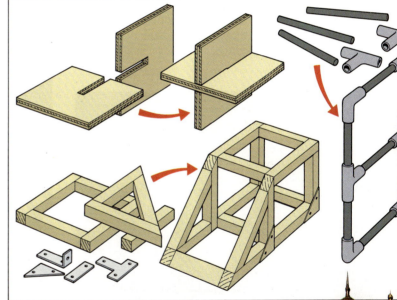

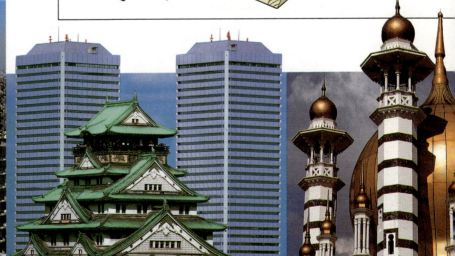

movement&
mechanisms

In any theatre presentation there is usually a lot of
movement on and around the stage. Obviously the
actresses and actors move about. There are also
scene changes, curtains rising and falling or opening
and shutting and backdrops changing. Sometimes
some of the props need to move around the stage.
Occasionally even parts of the stage have to move. In
some pantomimes or plays – *Peter Pan* by J.M. Barrie
or *Matilda* by Roald Dahl – the characters themselves
have to fly across the stage. In some productions –
mystery plays or ghost stories – there are special
effects, such as windows or doors opening by
themselves. All this movement has to take place
safely and in a controlled way.

·D A T A   F I L E·

Movement
Modelling
Ideas for modelling

You will need some of the following equipment for the activities on the following pages: construction kits (Meccano, Fischertechnic, Polymek etc), paper, wooden and plastic strips, paper clips, paper fasteners and drawing pins, gear and pulley kits, glue, and scissors or a knife.

levers & linkages

Look at some of the ways of getting things to move. Start by looking at some simple mechanisms: levers and linkages. Make up the simple linkages shown on this page and decide what type(s) of movement they produce.

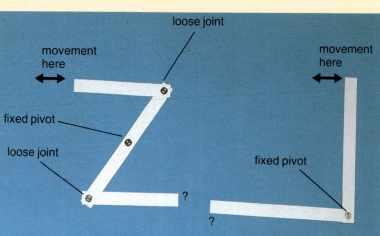

movement here

loose joint

fixed pivot

loose joint

?

movement here

fixed pivot

?

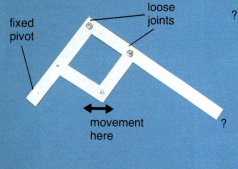

fixed pivot

loose joints

movement here

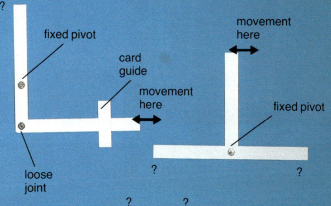

?

fixed pivot

card guide

movement here

loose joint

?

movement here

fixed pivot

?

?

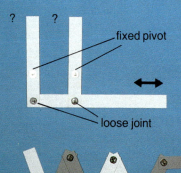

?   ?

fixed pivot

loose joint

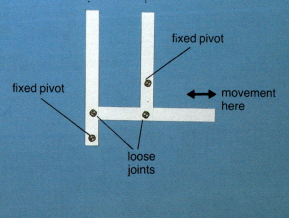

?   ?

fixed pivot

fixed pivot

movement here

loose joints

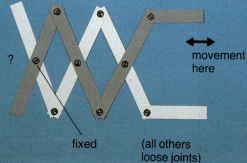

?

movement here

fixed

(all others loose joints)

· D A T A   F I L E ·

Movement Modelling

Use a piece of pin-board or a sandwich of corrugated card to model the linkages.

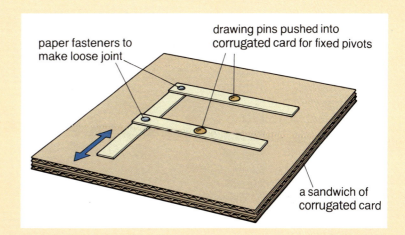

paper fasteners to make loose joint

drawing pins pushed into corrugated card for fixed pivots

a sandwich of corrugated card

Keep a record of your investigations in a table like this:

Where do you think mechanisms like these could be used on or around a stage?

linear	rotary	oscillating	reciprocal
(in a straight line)	(around in circles)	(rocking motion like a swing)	(backwards/forwards in a straight line)

A more complicated mechanism which uses linkages is called a **crank and slider.** Build this mechanism and investigate its movement. Keep a record of what you find out.

loose joint

loose joint

fixed pivot

slider

crank

What sort of movement do you think this type of mechanism would produce?

Where do you think mechanisms like these two could be used on or around a stage?

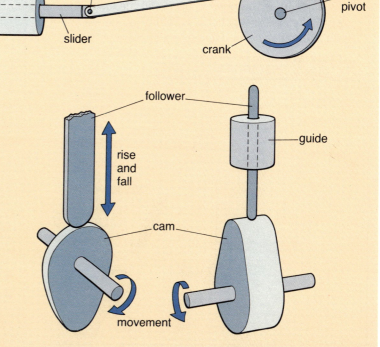

follower

guide

rise and fall

cam

movement

pulleys

A **pulley** system is another mechanism that can be used to produce movement. There are different types of pulley systems, from simple arrangements to complicated ones. Pulley systems can be used to open and close curtains, raise and lower scenery or move people and objects across the stage. Here are some different pulley systems. You might like to find out about other pulley systems

Build some pulley systems using a kit or from separate items.

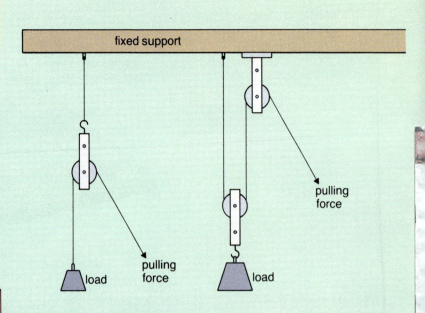

Use the pulley systems you have made to lift a weight or load. Use a force-meter or a Newton-meter to find out how much force you need to use to lift the load. Keep a record of your investigations. What do you notice about the force needed with different pulley systems?

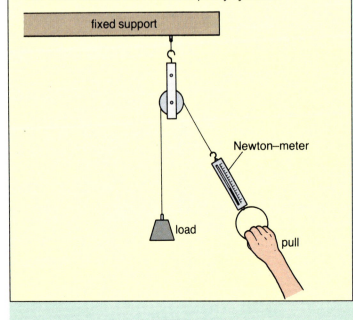

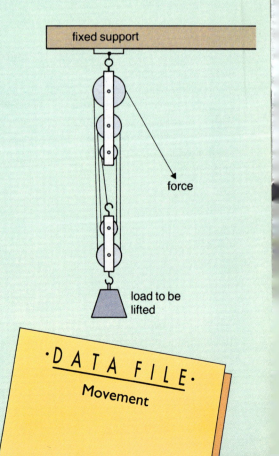

·D A T A   F I L E·

Movement

Here are some problems that
may be met in the theatre.

During a performance the curtains
obviously have to be opened and closed or
raised and lowered. Can you design and
build a mechanism for doing this using a
pulley system?

In *Peter Pan*, many of the characters
'fly'. They move up and down in the
air as well as across the stage. Design
and build a suitable mechanism to
produce this effect.

During a performance of *Mary Poppins*, Mary,
Bert and the children are gently lowered from
a tall building to the ground. Would a pulley
system or some other mechanism be suitable
to produce this effect? Design and build such
a system.

gears

In the past, and even in some theatres today, theatre staff pull the strings or wires that make movement happen. Often the theatre staff are helped by having counterweights attached to the wires. This means that they don't have to lift the entire weight themselves.

Sometimes it would be easier to use a motor to lift a heavy weight. However, the motor's shaft can spin very quickly so its rotation needs to be slowed down or controlled. Gears are used to slow down rotary motion. They can also be used to speed up rotary motion, change the direction or angle of rotary motion, or transfer rotary motion from one place to another. When several gears are used together they are called a **gear train.**

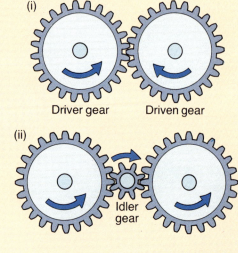

(i) Driver gear   Driven gear

(ii) Idler gear

(iii) 40 teeth
Changing speed   8 teeth

bevel gear   Changing direction

(iv)

Changing speed and direction   worm gear

worm wheel   (v)

rack and pinion

(vi)

Try producing some of these gearing effects with the materials you have. Keep a record of the type of motion you get with the arrangements you try.

Typical gear train in toy motor

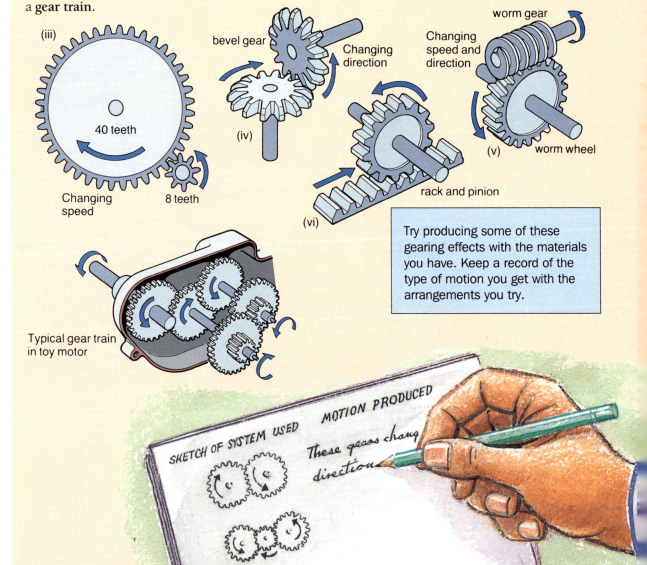

SKETCH OF SYSTEM USED   MOTION PRODUCED

These gears chang direction

A theatrical presentation involves the rotation of part of the stage. This part of the stage needs to be rotated fairly slowly. The initial design sketch by the set designers and the production crew looks like this:

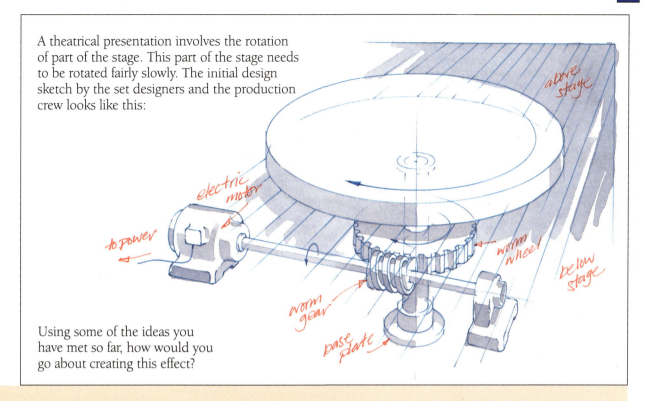

Using some of the ideas you have met so far, how would you go about creating this effect?

Here are some situations where large-scale movement is used in the theatre.

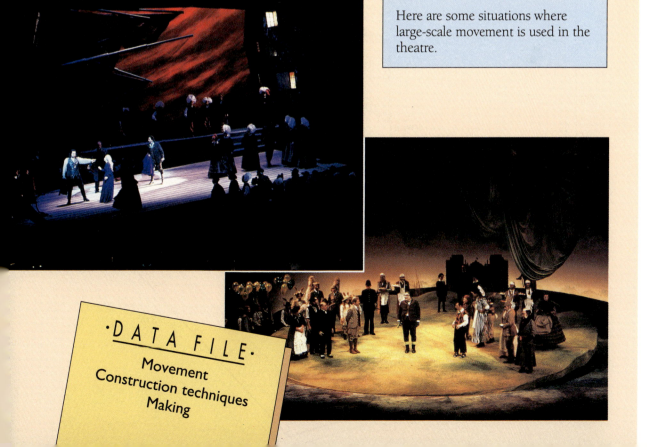

·D A T A F I L E·
Movement
Construction techniques
Making

a travelling show

Dolly and Guy Stroller earn a living by running a small puppet theatre company specialising in birthday parties for young children. They call themselves 'Guy's and Doll's'. They are usually booked to perform at parties held in church or community halls. They will also perform in the garden of a house where a party is being held. They need quite a bit of space to set up their stage. Occasionally they are invited to primary schools to perform in front of the whole school.

Sometimes they go to secondary schools to run creative workshops for the pupils. At these workshops they show the pupils how the puppets are made. They also describe how they manage to design their puppets to make quite complicated movements.

This is what their stage looks like:

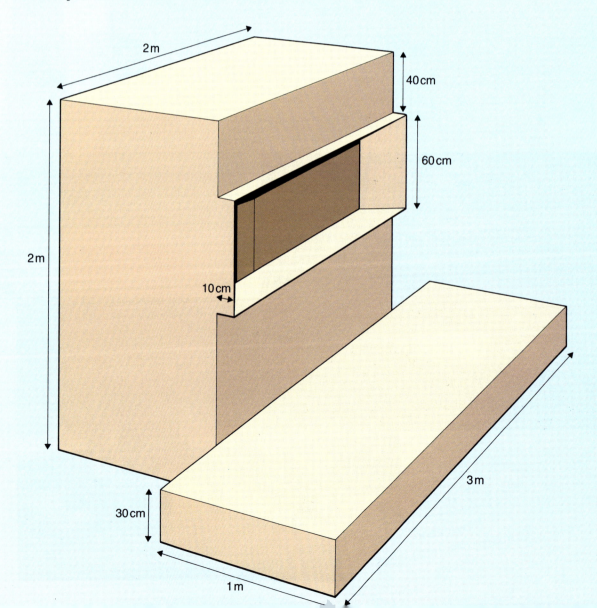

There are many different ideas to do with movement in this section. Your teacher may also have shown you other mechanisms that will make or change movement.

During one workshop in a school, Guy and Dolly told the pupils about their ideas for their next production. They wanted to try and get the following movements into their puppets:

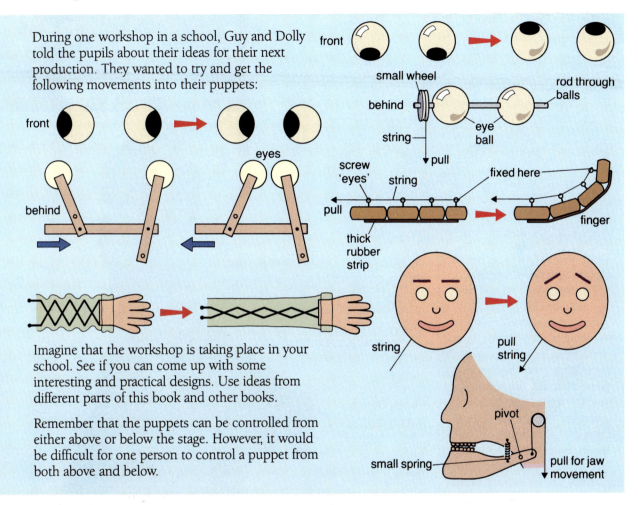

Imagine that the workshop is taking place in your school. See if you can come up with some interesting and practical designs. Use ideas from different parts of this book and other books.

Remember that the puppets can be controlled from either above or below the stage. However, it would be difficult for one person to control a puppet from both above and below.

Some other areas within the theatre where you might find opportunities for developing design and technology are:

- understage structure
- raised stages
- sound baffling, absorption or scattering
- overhead gantries to support beams for lights
- modelling scenes
- model of stage
- designing floor plans
- make-up
- recording ticket sales – using spreadsheets to update weekly or daily sales
- using a database to compile a mailing list
- using desk-top publishing to produce promotional material

·D A T A F I L E·
Construction techniques
Making

DRESS SENSE

There is a range of different activities in this section of the book. All of these activities or projects are related to some aspect of producing the costumes for a theatrical, television or film production. Your teacher can give you a clearer picture of how the various parts of the section fit together. Some pages contain clearly defined tasks, on other pages there are some fairly short activities, and on others there are opportunities for longer projects.

Costumes are a very important part of many productions and performances. It is important, for example, that the characters in a play set in modern times are dressed correctly. This might not be too difficult to achieve. The costume designers for the performance could buy clothes in any large department store. Some these clothes might need changing but probably not too much would need to be done.

It is even more important that the characters in a 'period' play are dressed correctly. Actresses wearing jeans would look out of place in a play set in Elizabethan times. Actors wearing armour would look out of place in a western. The costumes are usually designed to match the setting of the performance.

In this section of the book, you are going to be looking at the work of people concerned with the designing and making of costumes. This could be in theatre, film or television. You may be taking on the role of a costume designer, a costumes manager or a costumes assistant, to name just a few. You may already know something about the roles of these people and how they work together. You might like to think about the following situations.

Every year the town of Oldbury has a festival. On the final day of the festival, many town groups arrange a fancy dress carnival. As part of the celebrations a procession through the town is held. You belong to an amateur dramatic group which likes to make bright carnival clothes each year as part of the festivities.

Decide on a fancy dress theme and produce a range of stunning outfits for the gala.

School Films Ltd. are about to make a film of *Joseph and the Amazing Technicolour Dreamcoat*. Think about what the main characters will be wearing?

· D A T A F I L E ·

Sketching

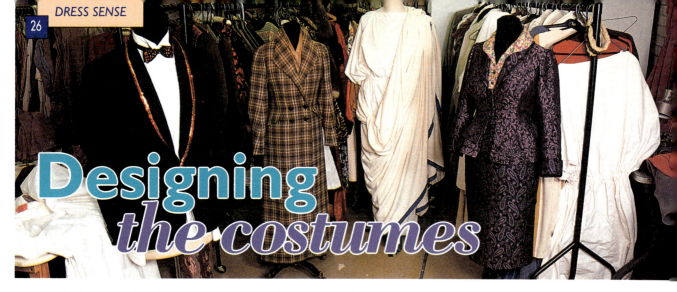

Designing
the costumes

The designing and making of costumes involves many different people. You could begin by looking more closely at what happens in a theatre, television or film costume department.

Most costume departments have a store of hundreds of different costumes. These costumes have been bought or made specially for particular performances. They may have been collected over many years. Sometimes an 'off-the-peg' costume from this store will be used for a performance. Sometimes a brand-new costume has to be designed and made.

Making
design sketches

Start by thinking about what a costume designer does. Designs for the costumes for a new performance have to be sketched out. The costumes could be for:

a science fiction serial on the television. The series is set in the year 4091. The action takes place on an airless planet many light-years away from Earth.

the theatre company in a town near you. They are planning to perform the musical *Oliver*. The story is set in Victorian England in the 1870s.

a film company which intends to make a version of *The Iron Man* by Ted Hughes. The film will be set in modern times.

You might like to think of a situation yourself. Try to choose a situation that would give you the opportunity to make interesting designs.

Your first job as a costume designer is to come up with interesting and original designs for the clothes that the characters will be wearing. These designs may only be rough sketches at first. Often the costume designer doesn't know what the finished product will look like.

The costume designer may write suggestions about how the costume should appear beside the sketches.

I WANT THE MAN'S COSTUME TO REFLECT THE CHARACTER HE IS PLAYING IN THE FILM

I WANT SOMETHING TO GIVE THIS A SHIMMERING EFFECT

THESE CHILDREN LIVED IN HARD TIMES AND THEIR CLOTHES NEED TO HAVE THAT GRUBBY LOOK ABOUT THEM

HERE ARE SOME ROUGH SKETCHES OF COSTUMES FOR A PANTOMIME

·D A T A   F I L E·

Sketching
Making a time plan

Designing *costumes on computer*

The sketches are usually done by hand. However, it is also possible to use a computer drawing package to sketch some quick ideas for costume designs. You may even already have used the computer to make some sketches and saved them on a computer disc. You could use these sketches as starting points. Just modify some of your earlier designs on the computer screen.

Thinking *about fabrics*

Once you have sketched out your design ideas you may want to think about the colour and type of fabrics to be used. You may be even able to suggest a pattern for the fabric.

Costing
THE DESIGNS

Next, the sketches would be passed to the costumes manager. The job of the costumes manager is to turn these sketches into real clothes. But before any work is done on the clothes they have to be costed. As the costumes manager, you will have to know:

- what types of fabric you need to buy,
- how much of each type of fabric you need to buy,
- whether you also need to buy other items such as:

hats	wigs
gloves	false beards and moustaches
scarves or shawls	shoes
ties	socks, tights and stockings
handbags	buttons
belts and braces	jewellery

Your next task will be to cost all the items needed.

·D A T A   F I L E·

Materials
Information systems

Keeping *within a budget*

The theatre manager, or the television or film production manager will have given you a budget to work from. Your budget will be the amount of money that you will be allowed to spend. You must make or buy the costumes within this budget.

One of the easiest ways of keeping track of the cost of items is to use a spreadsheet on a computer.

Spreadsheets are an excellent way of playing with numbers. You can experiment with different costs of materials. You can experiment with different lengths of materials. You can quickly see the result of buying hundreds, or thousands of buttons. If the cost of a fabric goes up you can immediately see the effect it would have on your budget. This will help you if you have two or three different designs for a costume. You will be able to see which would would be the cheapest.

Expensive dress planned...

Cambridge Theatre Group Costume Department

Pantomime: Mother Goose
Budget forecast for: Fairy Godmother (1 costume only)

Budget : £40.00   **Budget set on:** 01/11/91

Item	quantity	units	cost	total cost
cotton	5	metre	0.75	3.75
lace	10	metre	1.95	19.50
netting	2	metre	0.50	1.00
dowel	1	metre	0.33	0.33
feather boa	4	each	1.45	5.80
crepe de chine	2	metre	2.46	4.92
polyester	5	metre	1.69	8.45
elastic	2	metre	0.20	0.40
wire	6	metre	0.35	2.10
glitter	1	pack	1.25	1.25
			Total	47.50

Over (–) or under (+) budget forecast :-   −7.50

On these actual costings, the Fairy Godmother's outfit comes out as £7.50 over-budget. This may be overcome by either reducing the cost of this costume or reducing costs for some of the others.

Type the numbers from the spreadsheets on this page into your spreadsheet program. Alter some of the numbers and see the effect this has. For instance, you could change the length of material you buy or you could look at cheaper material.

·DATA FILE·
Information systems
Materials

Here is a typical spreadsheet produced by a costumes manager. The working budget is shown at the top. Different fabrics and different costs are shown underneath. The total cost of the costumes is also indicated. The figure at the bottom of the last column shows whether you are within budget (+),or overspent (–).

Your budget plan will help you to reach some decisions:

● which costume designs will be used,

● what materials you can afford,

● how much of each material you can afford.

Sometimes you may find that you cannot keep within the budget allowed. You will need to go back to your manager and ask for more money. This will mean someone else in the production getting less money! Think about what effect this might have on the production.

This spreadsheet shows that the estimated costs for all the costumes was £30.50 over-budget.

Can you identify which costumes turned out to be even more expensive than the estimate?

A decision was made to accept the increased cost of the Fairy Godmother's and Mother Goose's costumes. In order to stay within the budget savings had to be made elsewhere.

Can you identify where the two biggest savings have been made?

Cambridge Theatre Group Costume Department

Budget forecast for :   **Pantomime :** Mother Goose

Budget : £1000.00   **Budget set on :** 01/11/91

Item		number needed	estimated cost for each costume	actual cost for each costume	estimated total	actual total
Dresses:						
Mother Goose	costume 1	1	45.00	49.25	45.00	49.25
	costume 2	1	49.50	47.50	49.50	47.50
	costume 3	1	50.00	53.80	50.00	53.80
	costume 4	1	37.50	37.50	37.50	37.50
	costume 5	1	42.00	41.85	42.00	41.85
	costume 6	1	47.50	38.60	47.50	38.60
Jill	costume 1	1	39.00	42.75	39.00	42.75
	costume 2	1	37.50	45.50	37.50	45.50
Fairy godmother		1	40.00	47.50	40.00	47.50
Wicked witch		1	32.00	35.00	32.00	35.00
Suits:						
Jack, the principal boy		1	45.00	37.34	45.00	37.34
Simple Simon		1	37.50	30.50	37.50	30.50
Sir Cyril		1	48.00	45.87	48.00	45.87
Shirts etc:						
Jack, the principal boy		1	5.00	4.50	5.00	4.50
Simple Simon		1	5.00	5.35	5.00	5.35
Sir Cyril		1	5.00	5.50	5.00	5.50
Complete outfits:						
Priscilla, the goose		1	45.00	47.50	45.00	47.50
Dog		1	56.00	52.95	56.00	52.95
villagers		4	25.00	22.54	100.00	90.16
woodland creatures		6	20.00	15.00	120.00	90.00
Accessories:						
hats		6	4.00	5.50	24.00	33.00
gloves		5	6.00	5.50	30.00	27.50
shoes		16	2.50	1.50	40.00	24.00
undergarments		10	3.00	2.75	30.00	27.50
other items		–	–	–	20.00	20.00
Total expenditure :					1030.50	980.92
Over (–) or under (+) budget forecast:					–30.50	+19.08

...but may need to think about a cheaper alternative.

MAKING A
mock-up

The costumes manager and costume assistants now begin the work of turning the designs into reality. One of the first things they might do is to make a mock-up of the dress or costume.

There are two ways forward here. One way is to produce a full-size mock-up or toile. To do this, you will need a tailor's dummy or some other full-size model to work on. You could use a member of your group instead of a dummy and pin the toile on them. The other way is to use a scale model and produce a smaller version of the costume.

MAKING
a toile

A toile is a mock-up of the costume design made from fairly cheap material such as calico or muslin. You could also make a toile with stiff paper. A toile will allow you to see how the costume can be put together. When you put your toile together you should be able to work out where any pleats, darts or tucks should be placed. You should also mark on each part of your toile the 'run' or 'grain' of the material you intend to use.

Try and produce a complete toile for your costume design. You could add decorations at this stage. These could also be mock-ups. You will need to work from real body measurements. These measurements should be taken from the person for whom the costume is being made. The body measurements that are important for costume design are shown here:

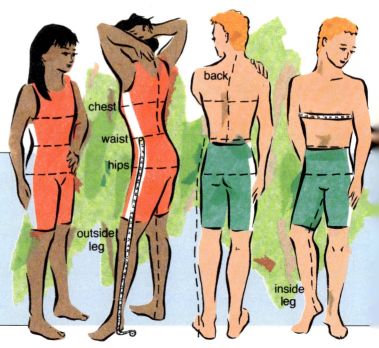

If the fabric or paper that you use to make your toile is held together with pins, be careful!

If you you use paper to make your toile then you will be able to produce the pattern for the costume very easily. Take the paper toile off your dummy very carefully, without tearing the paper. Gently flatten the paper out and cut away an surplus. Be careful not to cut away any paper you might need for overlapping seams or hems.

Now you have produced your basic paper pattern, you are ready to go ahead and make the costume out of real material.

·D A T A   F I L E·

Materials
Making templates

MAKING A
smaller version

It might, however, be very difficult or costly for you to make a full-size version of your costume. You may prefer to make a scaled-down model of the costume. Try not to make the model of your costume less than one-fifth normal size. If you make your model too small then you will run into problems to do with miniaturisation. For instance:

- making can get very fiddly if you are dealing with small parts.
- the kind of stitching used in the model might have to be different from the stitching used in the real costume.
- it might be difficult to miniaturise some of the decorative effects.

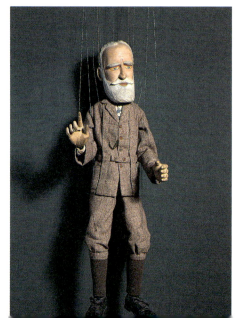

Making the pattern for your scaled-down model should be done in the same way as explained earlier: first make a toile on a scaled-down dummy and then produce a paper or fabric pattern from the toile. Maybe you have made a full-size toile but cannot make the full-size costume. In this case you will need to miniaturise the toile pattern. If you do this, make sure that the proportions of your finished pattern are the same as, or very close to, human proportions.

CREATING PATTERNS
on computer

You may have a computer drawing
package that allows you to save drawings
of individual parts of costumes as separate
computer files. You could then create and
store a whole range of patterns for the
separate parts of particular pieces of
clothing. Creating new patterns could be
done by bringing together the different
parts and modifying the result on screen.
You should be able to print, or plot, the
resulting pattern directly. The design may
be expanded or shrunk on screen before
being printed out. If you have such a
system then your designs can be 'tailor-
made' for your model.

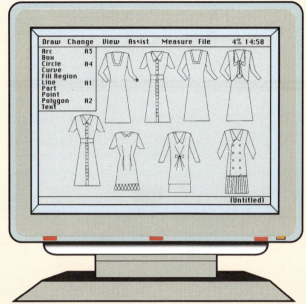

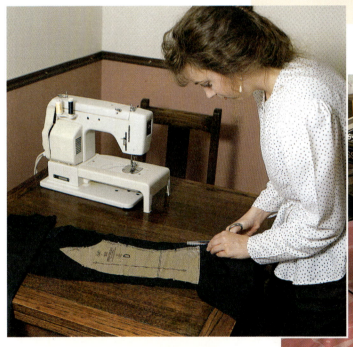

·DATA FILE·
Materials
Making templates
Information systems

Back to working
FULL-SIZE

Perhaps you don't want to make costumes from scratch. Sometimes it may be possible to modify old garments. You could make new costumes from old clothes.

You may need to add decorative effects to your costume. Many eye-catching effects can be achieved by adding simple decorations. You could use:

  lace
  braid
  fur collars and cuffs
  down or feathers
  appliqué
  bows or ties
  beads or sequins

These simple ideas can produce quite stunning effects.
  These extra decorations will need to be sewn or attached to the costume. Your teacher will show you a variety of sewing and fastening methods.

OLD SUIT

TROUSERS AND SLEEVES ARE SHORTENED

NEW COSTUME

·D A T A   F I L E·
Joining
Joints

Each part of a costume can be made in a number of possible styles. Sleeves, collars, cuffs, bodices, trousers, skirts and even coats can all be made in several different styles. The pictures here show some of the different forms in which the parts of a costume can be made. You might like to try using some of these ideas in your own costume designs. Your teacher may be able to show you different ways of making some of these designs. You will also need to know how they can be stitched together.

Of course, once you have made the costume, the actress or actor will have to try it on. The fit may not be perfect. You may have to make some small alterations to the design or to the shape or size. If you need help with how to do this, ask your teacher or look in a good sewing book which should show how to make many different kinds of adjustments.

What about all the extra bits and pieces that go with a costume. The costumes manager may have to make:

hats shoes jewellery
belts bags

ADDED
Extras

Obviously not all of these are made from scratch. The shoes needed for a performance would probably come out of the costumes department's stock. They may only need altering slightly. They may, for example, only need some decoration added to them.

Belts and bags are also usually taken out of stock and, if necessary, decorated in some way.

JEWELLERY

The costumes department may, however, need to make jewellery for some of the actors and actresses. Real jewellery would never be used. Coloured glass or beads, or shiny metal or plastic, would be used instead. Don't forget that the performers will be quite some way from any audience. The audience will not notice whether parts of the costume are home-made.

Here are some simple ideas for producing those little extras.

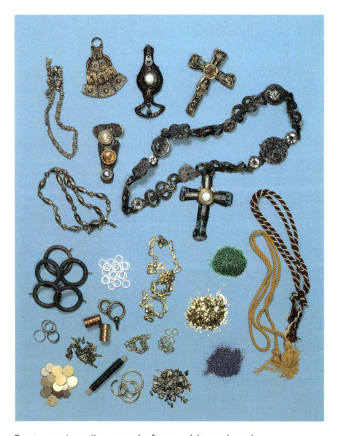

Costume jewellery made from odds and ends

HATS

Making hats for a performance is something that the costumes manager may also have to do. Hats, too, may be made very simply. Some kind of stiff card can be used as the base for many different designs of hats.

Papier maché may be used where you need a close fit over part of the head. Alternatively you could use a skull cap or some kind of net base.

And so the curtain goes up on yet another stunning, brilliant, dazzling, spectacular performance. And the critics loved it!

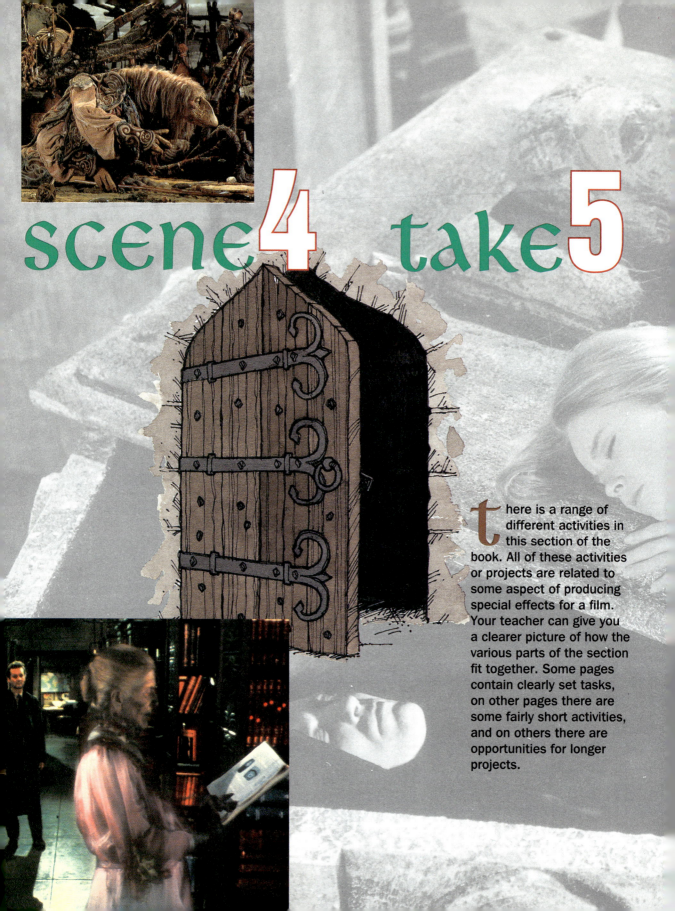

scene4 take5

there is a range of different activities in this section of the book. All of these activities or projects are related to some aspect of producing special effects for a film. Your teacher can give you a clearer picture of how the various parts of the section fit together. Some pages contain clearly set tasks, on other pages there are some fairly short activities, and on others there are opportunities for longer projects.

Special effects are used in different types of films – drama, adventure, horror or science fiction, for example. In this section you will be looking more closely at the use of special effects within two of these – horror and adventure films. Special effects are needed to create illusions, ghosts moving or skulls talking, scenery or props moving in a creepy way or weird noises. These special effects can often be made to happen using pneumatics or hydraulic systems. In this section you will be using both of these types of systems to create your own special effects

The first thing you will do in this section is to look at some simple pneumatic systems. Pneumatics is all about using air to do something useful for you. Simple pneumatic systems can be very useful and produce some interesting effects.

·DATA FILE·

Systems
Movement

Simple **pneumatic** systems

P neumatics is about using air to do work for you. You use air when you blow up a balloon or a car tyre. In both of these cases air is squashed or **compressed** into an enclosed space. If you blow up a balloon and then let it go or open a car tyre valve, the air rushes out. Compressed air like this can be made to do things for you.

Here are some other examples of pneumatics or compressed air being used to do something useful. See if you can spot where the compressed air is being used.

Pneumatic systems also provide a very useful way of transferring movement from one place to another. This is the same as saying that a pneumatic system will transfer energy from one place to another.

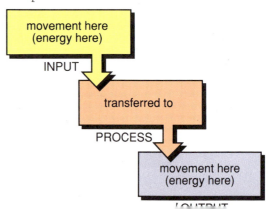

movement here
(energy here)

INPUT

transferred to

PROCESS

movement here
(energy here)

You may already be familiar with using plastic syringes. They can be used to make simple pneumatic systems.

piston rod

piston

You will need some of the following equipment for the activities on this page: syringes, plastic tubing, scissors, stiff card or small section timber, elastic bands, glue, balloons.

Make this simple pneumatic system.

Make sure that you can get as much movement as possible when you push and pull the pistons in and out. Remember, pneumatics is about using air. Pneumatic systems use air to transfer movement (or energy) from one place to another.

Here are a couple of ideas for you to try, using this simple system. Try to think of other ideas that might fit into the setting of a horror or adventure film.

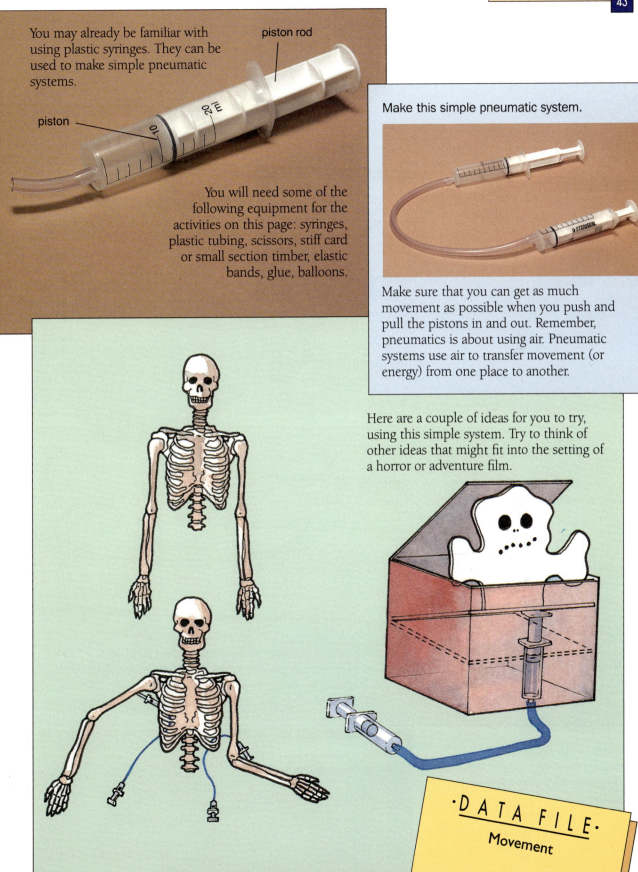

·DATA FILE·
Movement

simple hydraulic systems

Using the simple pneumatic system you have made, get someone to push in the piston of one of the syringes as firmly as possible. Push the piston of the other syringe into its cylinder, gently at first and then more and mor firmly. What do you feel? Now try to lift something heavy using your pneumatic system. Air is very squashy! If the weight you are trying to lift is too heavy, the pneumatic system will not move it. You will only succeed in squashing the air in the system.

hand pushing here

heavy weight

air

Fill the system with water using one of these methods. You could add a few drops of food colouring to the water; you will be able to see the water move then.

water

immerse in water

water

Make sure that there is no air inside the tubing. Get someone to hold the pistons of one of the syringes as firmly as possible. Push the other piston into its cylinder gently at first and then more and more firmly. What do you feel?

Water is not squashy, it is **incompressible**. Using water in a system turns it into a **hydraulic** system.

Once one of the pistons has been pushed in, you can withdraw it again to reverse the action.

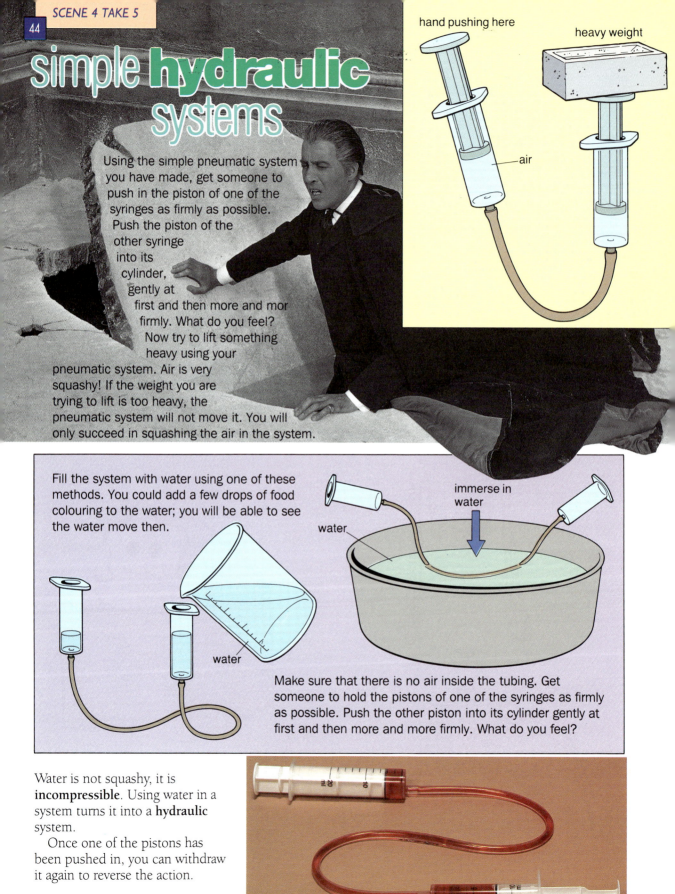

Pushing and pulling the pistons can be a bit of a nuisance. It would be better if you could move the pistons backwards and forwards mechanically. You will need some windscreen washer pumps. These may be available from scrap yards (cheap), school equipment suppliers (reasonably priced) or motor accessory shops (expensive).

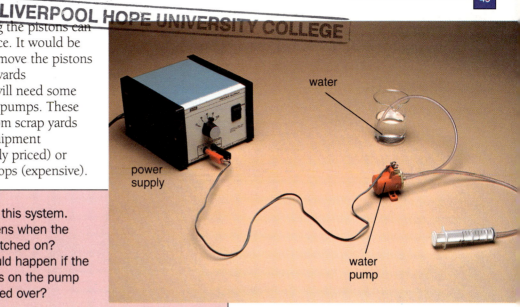

water

power supply

water pump

Connect up this system. What happens when the pump is switched on?
What would happen if the plastic tubes on the pump were swapped over?

Try a reversing switch in the power line.

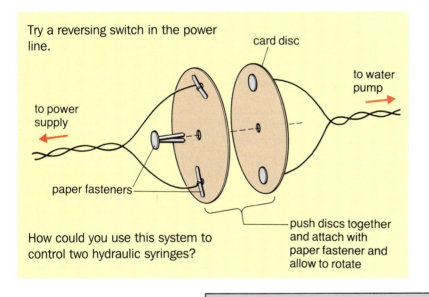

card disc

to water pump

to power supply

paper fasteners

push discs together and attach with paper fastener and allow to rotate

How could you use this system to control two hydraulic syringes?

Here is a useful electronic system that you might like to build. With this system you can do two things:
● switch the power to the pump on and off,
● reverse the direction of the water through the pump.

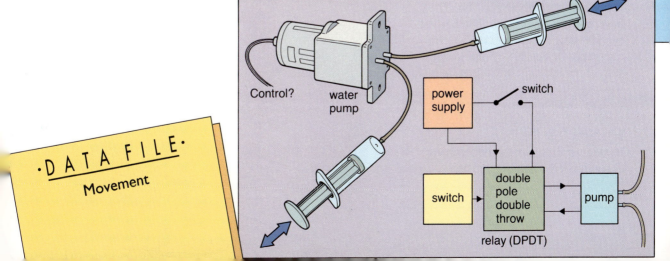

Control? water pump power supply switch

switch double pole double throw pump

relay (DPDT)

·DATA FILE·

Movement

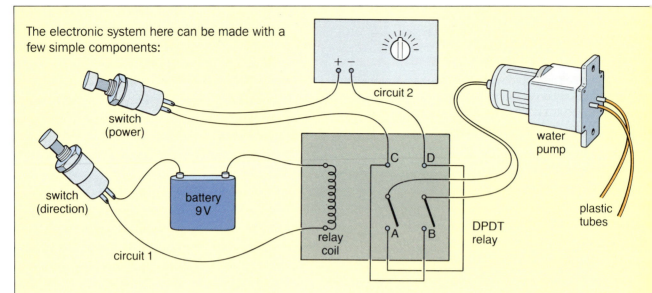

The electronic system here can be made with a few simple components:

switch (power)

circuit 2

water pump

switch (direction)

battery 9 V

relay coil

C D

A B

DPDT relay

plastic tubes

circuit 1

When circuit 1 is connected, the relay contacts are set to A and B. This means that the electric current in circuit 2 flows through the water pump in one direction. (Follow the current from the power supply, through the contacts, and back to the power supply.) When the direction switch is pressed, the relay coil is energised.

The relay contacts then jump to C and D. The electric current in circuit 2 now flows through the water pump in the other direction. (Follow the current from the power supply, through the contacts, and back to the power supply again.) Convinced?

Using plastic syringes should have given you some ideas about using pneumatics and simple hydraulics. You should be able to think of some interesting ways of using these within the context of this chapter of the book – horror and adventure films.

Try this for size.

Virginia Pictures Incorporated are making a spoof adventure film. They have called it *Ohio Smith*. The hero of the film gets into all sorts of escapades but usually escapes by the skin of his teeth. In one scene he is walking through a narrow passageway. He stumbles over a trip-wire which causes a dragon's head to fall down. The head moves from side to side warning him of the terrifying consequences should he go any further.

Using real
pneumatic
cylinders

Controlling the flow of air in pneumatic cylinders is far less messy than controlling water. However there are several precautions that you need to take before you use real pneumatic cylinders in the classroom.

- Fit an on/off valve in the compressed air line.
- Keep air lines well away from ears and eyes.
- Never point an air line at yourself or anyone else.
- Wear protective goggles.
- Make sure any open cuts on your skin are covered up.
- Make sure hands are kept well away from anything that's moving.
- Do not switch on until all pipe work is secure and all connections are made.
- Turn off the air supply before making changes to the circuit.
- Get your pipe work checked before you switch on.

In this section of the book you will be using a variety of pneumatic devices. You will need to have available:

single-acting cylinders
double-acting cylinders
three-port spring
   return valves
three-port solenoid-
   operated valves

five-port valves
flow restrictors
shuttle valves
some electronic building
   blocks from an
   electronic systems kit

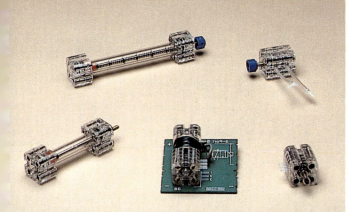

Some systems using real pneumatic cylinders

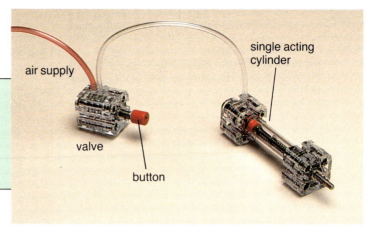

air supply

single acting cylinder

valve

button

Start by connecting up this system.

Press the button on the valve. What happens?
Release the button on the valve. What happens?

Not very exciting is it? Well, it could be put to lots of uses.

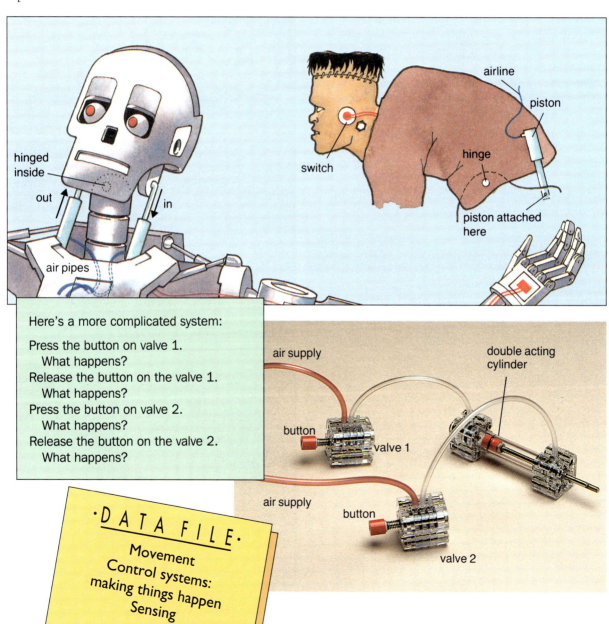

hinged inside

out

in

air pipes

switch

airline

piston

hinge

piston attached here

Here's a more complicated system:

Press the button on valve 1. What happens?
Release the button on the valve 1. What happens?
Press the button on valve 2. What happens?
Release the button on the valve 2. What happens?

air supply

double acting cylinder

button

valve 1

air supply

button

valve 2

·D A T A   F I L E·

Movement
Control systems:
making things happen
Sensing

Controlling the speed of pneumatic systems

In all of these systems the movements happen very quickly. You can slow down the speed of the pistons by placing one-way flow restrictors in the air line.

That's a bit more exciting isn't it? This simple system could be put to lots of uses.

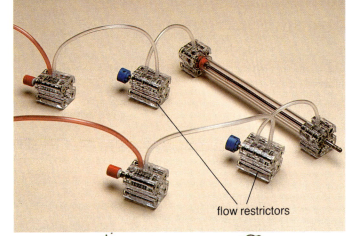

flow restrictors

So far you have seen how pneumatic cylinders can be controlled manually. The valves which direct the flow of air to the cylinders can also be controlled electronically. To do this you will need to use a solenoid-operated valve.

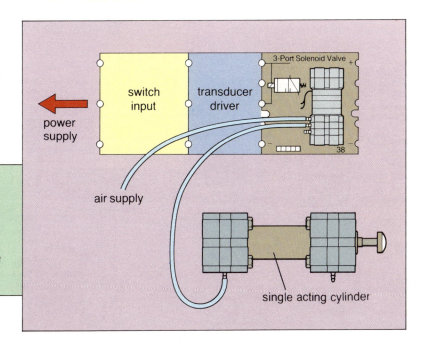

power supply

switch input

transducer driver

3-Port Solenoid Valve

38

air supply

single acting cylinder

Connect this system together:

Press the button on the switch input unit. What happens?

Replace the switch input unit with a light sensor. Cover the light sensor. What happens?

Using sensors to control pneumatic systems

You could use many different input sensor boards in this system. Try some of them out. Think of different uses for the system within the context of adventure and horror films.

You can also control a double-acting cylinder using solenoid-operated valves. You will need two three-port solenoid-operated valves for each double-acting cylinder that you use.

The two switch input units can also be replaced by other sensors in the electronics kit. Try two light sensors or one light sensor and one magnetic sensor.

A more economical way of controlling a double-acting cylinder would be to use a three-port pilot-spring return valve.

Again, think of different uses for these systems within the context of adventure and horror films.

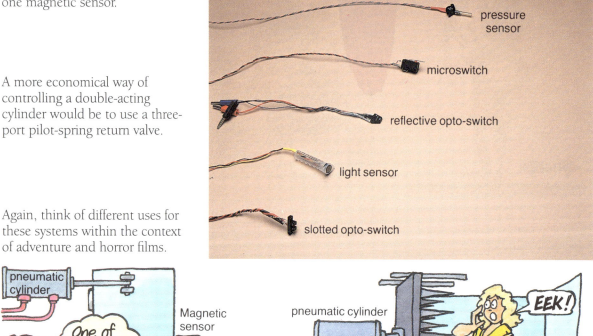

temperature sensor

pressure sensor

microswitch

reflective opto-switch

light sensor

slotted opto-switch

Many of the pneumatic systems that have been shown so far can be triggered by **remote sensors**. A sensor is anything which detects change and produces an electrical signal. Changes could be in light levels, movement, temperature or magnetic fields. Remote sensors may be used to control pneumatic systems. The sensors could be placed wherever you like. They are often placed far away (remote) from the pneumatic and electronic systems they control. For instance, a sensor could detect people walking near a door. A pneumatic system could then be switched on which would open the door. A sensor could detect the rise in temperature of a greenhouse. This could cause a pneumatic cylinder to open the greenhouse windows.

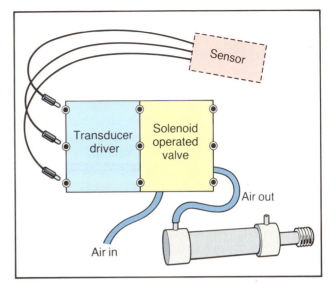

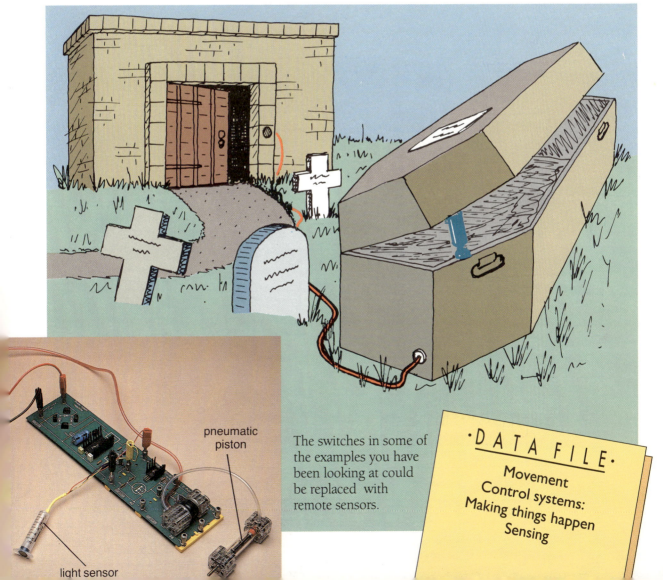

pneumatic piston

light sensor

The switches in some of the examples you have been looking at could be replaced with remote sensors.

·DATA FILE·
Movement
Control systems:
Making things happen
Sensing

Carefully California Kate crept down the passage. She edged her way through a tangled mass of rotten wooden beams, reeking of decay and corruption. She touched a delicately hanging overhead timber and sent a shivering, slurp of slime trickling down her neck. Her shoes felt the oozing dampness underfoot. Slowly, oh so slowly, she stretched out her hand towards the gleaming, glistening, glowing gem that sat motionless on a moss covered pedestal. Her fingers were poised to snatch the priceless jewel when she noticed the wire. She froze, motionless, with fear.

She had seen a trip-wire protecting the jewel. She knew something horrible would happen if she lifted it off the pedestal. All sorts of things went through her mind. What was to be her fate.

Here are some possibilities. Which one of these would make a good scene in a film. Try to model the situation.

You've had a good idea for a scene in a horror film called
The Haunted Hand. You explain to the director what will
happen.

> 66 In this scene the hand emerges from a cupboard
> as you enter the room. when you try to get too
> close to the hand it retreats back into the cupboard.
> when you move away, the hand comes out again. 99

The director likes the idea and asks you to make a
prototype of it.

You are part of a film company producing remakes of some of
the classic horror and adventure movies from the 1930s –
Dracula, *Frankenstein's monster*, *King Kong* and *Flash Gordon*.

Many of the special effects
for these films can be
achieved with simple
pneumatics. Take any one
of these films and think
about some scenes from it.
Try to come up with some
set or character designs
where pneumatic systems
would be useful.

·DATA FILE·

Materials
Control systems:
making things happen
Sensing

THE SCORES ON THE DOORS

There is a range of different activities in this section of the book. All of these activities are related to some aspect of television quiz shows. Your teacher can give you a clearer picture of how the various parts of the section fit together. Some pages contain clearly set tasks, on other pages there are some fairly short activities, and on others there are opportunities for longer projects.

There are lots of different TV quiz-shows – The Generation Game, Telly Addicts, Mastermind, The Krypton Factor, Blockbusters, and Every Second Counts, to name but a few. In most of these shows, people compete against each other for points. The contestants often have to press a button when they know the answer to a question. TV quiz-shows often need a score-board. The score-board is sometimes displayed in front of the contestants or there may be a very large display in front of both the contestants and the audience. In fact, the types of displays used in different programmes vary quite a lot.

The score-board could be:
● a digital display
● a dial with an arrow moving over it, or
● a series of boxes which light up or go out as the players' scores change.

Lots of electronic bits and pieces are used in making a TV show, both behind the scenes and in front of the TV cameras. There are electronically controlled lights, buttons and displays.

Electronics are also used to control lighting, sound and camera movement.

The lighting in a TV studio needs to be controlled. Spotlights need to shine on the main performers in the programme. The title music needs to begin and end on time. There may also be short pieces of music to link one part of the programme with another. Music may be needed to introduce different types of question. All of these effects have to be carefully designed to produce a slick and smooth-running programme.

Fingers on the buttons

On these two pages you will be looking at some electronic systems that might be used in TV programmes of this kind. You can imitate some of the effects used in TV quiz-shows with an electronic systems kit.

You will be designing and building systems which can be used to light lights or buzz buzzers. Electronic devices to work out which button was pressed first may be needed. An electronic scorer, may also be needed to keep track of the scores and display them.

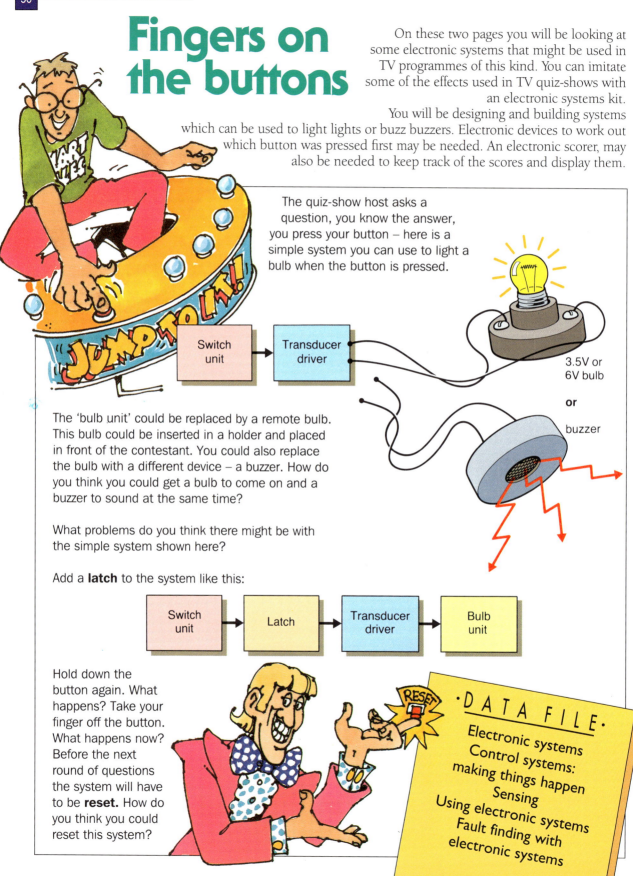

The quiz-show host asks a question, you know the answer, you press your button – here is a simple system you can use to light a bulb when the button is pressed.

Switch unit → Transducer driver

3.5V or 6V bulb

or

buzzer

The 'bulb unit' could be replaced by a remote bulb. This bulb could be inserted in a holder and placed in front of the contestant. You could also replace the bulb with a different device – a buzzer. How do you think you could get a bulb to come on and a buzzer to sound at the same time?

What problems do you think there might be with the simple system shown here?

Add a **latch** to the system like this:

Switch unit → Latch → Transducer driver → Bulb unit

Hold down the button again. What happens? Take your finger off the button. What happens now? Before the next round of questions the system will have to be **reset**. How do you think you could reset this system?

·D A T A F I L E·

Electronic systems
Control systems:
making things happen
Sensing
Using electronic systems
Fault finding with
electronic systems

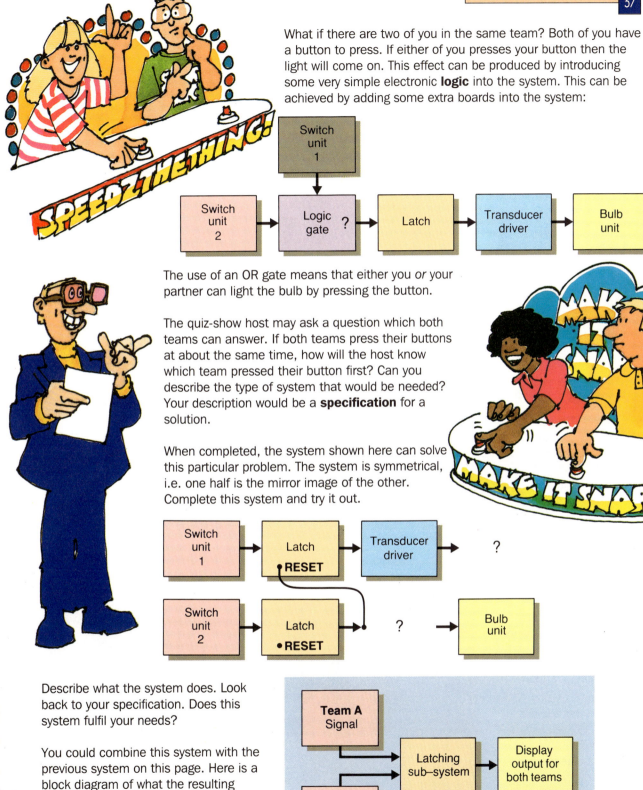

What if there are two of you in the same team? Both of you have a button to press. If either of you presses your button then the light will come on. This effect can be produced by introducing some very simple electronic **logic** into the system. This can be achieved by adding some extra boards into the system:

Switch unit 1

Switch unit 2 → Logic gate **?** → Latch → Transducer driver → Bulb unit

The use of an OR gate means that either you *or* your partner can light the bulb by pressing the button.

The quiz-show host may ask a question which both teams can answer. If both teams press their buttons at about the same time, how will the host know which team pressed their button first? Can you describe the type of system that would be needed? Your description would be a **specification** for a solution.

When completed, the system shown here can solve this particular problem. The system is symmetrical, i.e. one half is the mirror image of the other. Complete this system and try it out.

Switch unit 1 → Latch **RESET** → Transducer driver → **?**

Switch unit 2 → Latch **RESET** → **?** → Bulb unit

Describe what the system does. Look back to your specification. Does this system fulfil your needs?

You could combine this system with the previous system on this page. Here is a block diagram of what the resulting system would look like. Use an electronics system kit to build this system.

Team A Signal
Team B Signal → Latching sub–system → Display output for both teams
Questioner's resetting sub–system

Keeping score

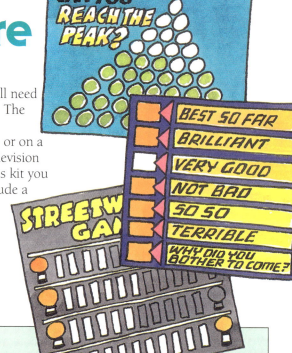

In any TV quiz-show a record will need to be kept of the players' scores. The scores might be displayed on contestants' desks or consoles, or on a separate score-board in the television studio. The electronic systems kit you have been using should include a counter. A counter of this type could be used to help the quiz-show host or the scorer to keep a record of the points awarded to the various contestants.

Connect this system together:

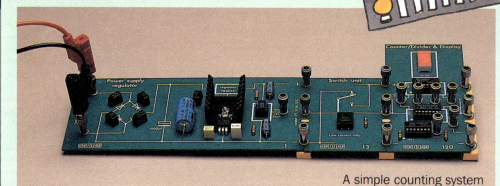

A simple counting system

Press the button once. Press the button several more times. What happens when the display shows the number nine and you press the button again? Try this. Make sure that the display resets itself to zero. If you want to count past nine you will need a second counter display. The second counter can be connected like this:

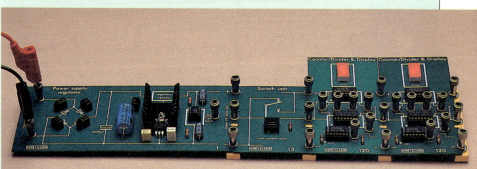

A system to count to 99

Take care!

The display will show the numbers in the wrong order – units and tens instead of tens and units – so when your counter has counted up to 12, your display will read 21.

In some quizzes you may want to know when a team or contestant has scored a certain number of points. You might, for instance, want a bell to ring when one team had scored six points. You would need to know how to make an electronic system 'recognise' that a certain score had been reached. This can be done using some very simple electronic logic.

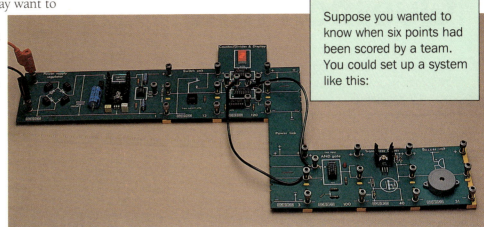

Suppose you wanted to know when six points had been scored by a team. You could set up a system like this:

This system can now be built into your larger system. You can in fact, use this system to detect any score, provided you have the correct logic.

Design an electronic system that would 'recognise' when five points had been scored. Does this system also allow you to detect when 12 points have been scored?

The quiz could end when one of the teams scores 25 points. When this happens you may want something dramatic to happen – a bell or buzzer could go off, lights could flash or a triumphant musical fanfare could sound.

Switch input → Counting sub–system → Display output

Counting sub–system → Detecting a score of 25 sub–system → Display output

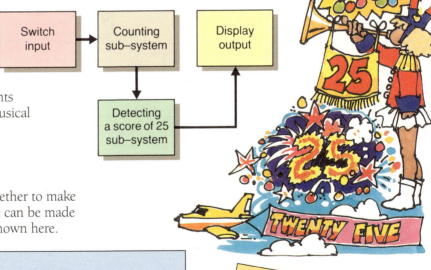

Here's a clue to help you:
logic gates can be grouped together to make larger units. A three-input gate can be made from two two-input gates as shown here.

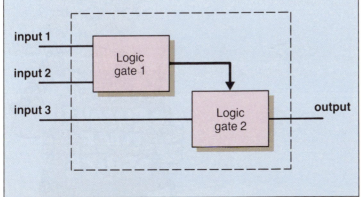

input 1, input 2 → Logic gate 1 → Logic gate 2 ← input 3 → output

· D A T A   F I L E ·

Electronic systems
Control systems: making things happen
Sensing
Using electronic systems
Fault finding with electronic systems

Question time

Contestants in quiz-shows are often asked to answer several different types of questions. They may be asked questions about sport, pop stars, TV programmes, general knowledge, natural history and dozens of other subjects. Often the contestants have a free choice of which type of question they want to answer. Sometimes the subject of the question is chosen for them, by the throw of a dice or the spin of a wheel for example.

There are also quiz-shows like Trivial Pursuit where you might have to shake a dice or spin a pointer. The question you are asked depends on where you land on the game board.

Making an electronic dice is not difficult. The electronic system you will need is shown here. The box marked 'logic gate' controls the pulses going to the counter. You will have to decide for yourself what logic gate you will need. Once the system is working, try to describe how it works. Think about the signals moving from one part of the system to another.

You could easily make this system more elaborate. You could, for instance, award an extra go, or extra points, if a particular number appeared on the dice and you could make the system sound a buzzer every time the number appeared.

· D A T A   F I L E ·

Electronic systems
Control systems:
making things happen
Sensing
Using electronic systems
Fault finding with
electronic systems

Time's Up!

Here's a little problem for you to try.

In TV quiz-shows the contestants often have only a short time to answer. Design an electronic system that will sound a buzzer when the time is up. You will need an electronic clock to do this; a pulse generator is an electronic clock. Make sure that your pulse generator 'ticks over' at a steady rate – e.g. one pulse per second or one pulse every two seconds.

You will need to think carefully about the problem before you start.

Think about:
- the different sub-systems you will need,
- the order in which these sub-systems should be arranged,
- the signals in different parts of the system,
- how these signals move through the system.

Editing the show

Most television programmes are recorded on video tape. After the whole programme has been recorded a lot of post-production work takes place. This is when the programme is edited into its final version for broadcasting. This might involve:

- adding title sequences at the beginning and end of the programme.
- adding any special effects during the programme.
- adding short, linking pieces of music.
- adding short, linking cartoons or other pieces of animation.
- removing unwanted sequences – e.g. clapper-boards, introductions.

Here are three problems related to post-production work. They all involve the use of logic gates.

1 A sound track needs to be added to a video tape. The sound track is on one machine, the video tape is on another. The two are to be superimposed on a third machine. The first two machines each send a signal to the third machine to say when they're ready. Design an electronic system to start recording on the third machine when the first two machines are ready.

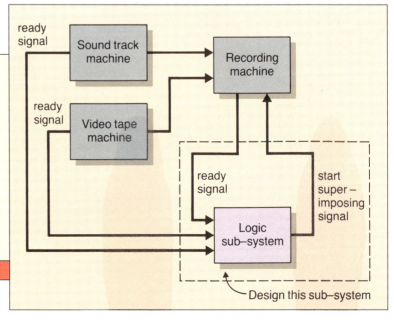

Design this sub–system

2 A piece of music has to be added to a video tape at a certain time: the piece of music must start six seconds into the programme recorded on the tape. Design a system that will send a signal to the video machine to start recording the music at the right time.

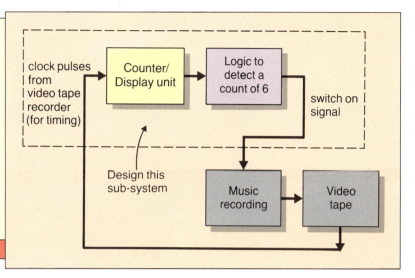

clock pulses from video tape recorder (for timing)

Counter/ Display unit

Logic to detect a count of 6

switch on signal

Design this sub-system

Music recording

Video tape

3 A piece of music linking two scenes in a TV programme has to be added to a video tape. The music must start three seconds into the programme. The music must stop nine seconds into the programme. Design a system that will send a signal to the video machine to start and stop the recording of the music at the right times.

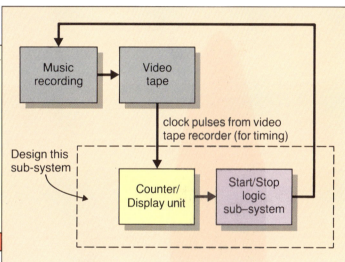

Music recording

Video tape

clock pulses from video tape recorder (for timing)

Design this sub-system

Counter/ Display unit

Start/Stop logic sub-system

·D A T A   F I L E·
Electronic systems
Control systems: making things happen
Sensing
Using electronic systems
Fault finding with electronic systems

Studio lights

Many TV quiz-shows have bright, colourful sets. There may be lots of lights going on and off. In some TV programmes you will see lights flashing to the beat of music. Interesting effects like this can be produced very easily.

You can produce lighting effects like this yourself. You will need the following building blocks from your systems kit:

Connect up the system like this:

Now try talking into the sound sensor. Try whistling. Try clapping. What happens to the bulb unit? The system can be made more sensitive by adjusting the preset dial on the comparator like this:

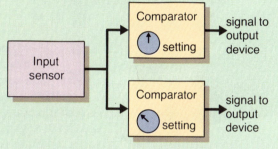

rotate this dial to adjust the sensitivity

Try singing into the sound sensor. Try playing a tape recording of a piece of music: use some music with a good strong beat. Watch what happens to the bulb unit.

To make it even more interesting you could use two comparators set to different levels of sensitivity.

The comparator is a very useful electronic device. Electronic sensors produce signals in response to changes in temperature, light level, and other measures.

A comparator can be used to detect when these signals go above or below a preset level.

A comparator could be used to make a safety cut-out for TV studio lights. If the lights became too hot, they could be automatically shut down. This would be quite complex to do, but an outline of a system which should work is shown here. The names of some of the electronic blocks are missing: you will have to work these out for yourself!

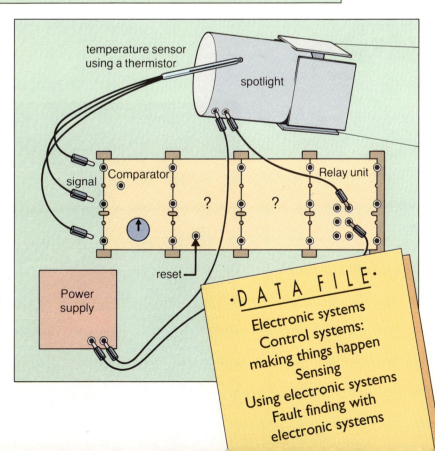

·D A T A F I L E·

Electronic systems
Control systems:
making things happen
Sensing
Using electronic systems
Fault finding with
electronic systems

SOUNDS LIKE...

There is a range of different activities in this section of the book. All of these activities are related to some aspect of making a radio show. Your teacher can give you a clearer picture of how the various parts of the section fit together. Some pages contain clearly set tasks, on other pages there are some fairly short activities, and on others there are opportunities for longer projects.

Take a look at the pictures on these two pages. They show how radio waves are sent through the air, how they are detected, and how they are converted into the sounds you listen to and other types of information. In this section you will be looking at how you can send and detect radio waves. You may even be able to use some very special equipment which will let you 'see' the radio waves. To start with, you will look at how you might go about making a radio programme.

Writing a radio show script

You are part of a group which is going to produce a half-hour music programme for a local radio station. This could be a local commercial station. You will be producing a script for your radio programme, using a word-processing package. One of the good things about word-processing is that you will be able to make changes to your script quickly and easily without having to re-type whole pages of notes and instructions.

Your group should start by thinking about what type of music programme they want to produce. There are lots of types of music to choose from: pop, classical, rock, country and western, rap, reggae or any other type of music you enjoy. You might choose to play a mixture of different types of music in your programme or you might like to produce a programme of music from other countries or cultures. Think about whether you might want to interview somebody on your programme. What questions would you ask them? Try to identify the type of things that will need to appear in your script. You will need to include:

what you are going to say

- your introduction to the programme
- who you are
- what your programme is about
- what you are going to talk about during the programme
- who your guest is, if you are having a guest
- your interview questions for your guest
- your closing words
- any advertisements

your choice of music

- the name of each piece of music and the names of the singers, bands or orchestras performing the music
- a few words about the performers and their music

your instructions for your studio crew

- who is going to do what
- when they are going to do it
- when jingles, adverts, etc., need to be played

any cues you will need to give or to watch or listen for

- visual cues to and from the studio crew
- sound cues from the beginning and end of each piece of music

You should now be ready to write out a rough outline of your script. Do this on paper to start with. You will probably think of all sorts of things which you have missed out as you go along. Don't worry. You will be able to insert these quite easily with the word-processor.

Once you have put together a rough outline of what you want to say, start typing it in at the keyboard of the word-processor. Don't worry about starting new lines – the computer will do that for you. You will have to decide where to start new paragraphs though!

Don't worry about making spelling mistakes, missing words out, or putting in incorrect punctuation. Just keep typing.

·D A T A   F I L E·
Information systems

Once you have finished typing your rough script into the computer, you can start to organise it in the way you want.

Suppose when you print out your rough script, it looks like this. You can see that there are lots of corrections or changes that need to be made. Compare this with your text.

```
Music at start of programme
Jingle                    —spelling
Hello and welcome to this programme of pop music on
Tuesday morning. This is Sandy Beech and Rocky Cove
bringing you half an hour of music, madness and
mayhem on your fab fun local radio station.
             Put my name here - Sandy           Spelling
need to write cue
Here we are off to a bright start with that new smash
hit from the Moon Monkeys — Can't stop aping around
you.


Well that was brill. Moon Monkeys are a new group
just starting out in the pop world and they deserve a
big break
          add music (jingle?)
Advertisement for clothes from local supplier here

                    Add in joke about the
                         long trousers — Sandy.
What's next Rocky whose popped into the studio today
to give a big wave to the listeners?
        Rocky says this put name
               in here
Well Sandy we've had lots of people write in and say
why dont we get a big name in to talk to us. Well
listners here for a chat is the heavy rock group Man
Mountain and he's brought his
backing group Twin Peaks with him.


Man Mountain how did you get into
the pop world
```

·DATA FILE·
Information systems

You can see that lots of changes will need to be made to this rough script. These changes will make the script easier to follow. They will make it clear who is speaking at each point, the cues for the studio crew, when to play jingles and adverts, and when to start talking over the music.

You will probably want to make changes of this kind to your script. You may also want to move text around and change the order of the words.

You may want to emphasise, or stress, certain words in your script. This could be done by making the words bold, putting them in italics, underlining them, or even changing the font or the size of words.
If several of you are going to read the script, you will need to decide who reads which part. The script must say clearly who each line belongs to. This can be done by indenting the text like this.

You could try shortening the length of the lines across the page. This might make the text easier to read. You should be able to move the text to the left or to the right. Try changing the left- or the right-hand margin. Try centring some text – your headings for instance.

There are many different types of word-processing package. Most of them do similar sorts of things, though some will have more features than others. Your word-processing package may be able to do some of the following:

– searching automatically for a particular bit of text and replacing it with new text.
– automatically writing special headings and footnotes on each page.
– automatically numbering pages.

Some word-processors also have a spell-check built in. As a final task, check through your spelling.

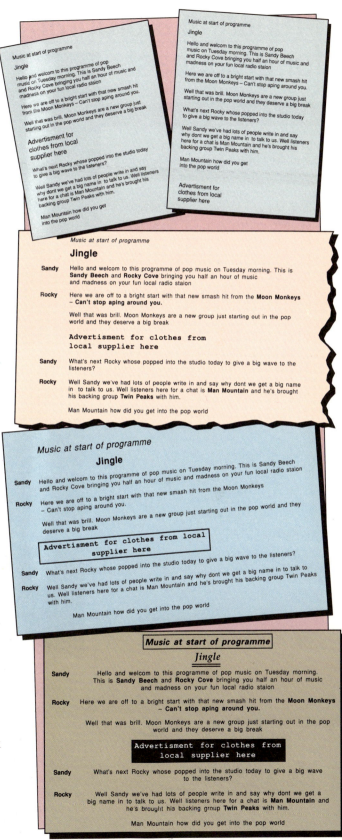

Preparing a programme for broadcast

AMPLIFYING THE SIGNAL

The electronic signals generated in the radio studio are tiny compared with the signals broadcast from the radio station's aerial. Any signal which is to be broadcast has to be amplified greatly first. On these two pages you will look more closely at how electronic amplifiers work and at some of their uses. This may sound difficult, but don't worry! If you follow the instructions here, you won't get bogged down in complicated electronics, and if you have an electronic systems kit, you will find the activities very straightforward.

Amplifiers are designed to make signals bigger. If you put a small signal in, you should get a larger signal out.

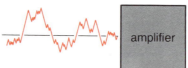

·D A T A   F I L E·

Using electronic systems
Fault finding with
electronic systems

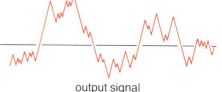

input signal   amplifier   output signal

Connect up one of these two systems together:

If you are using an oscilloscope, your teacher will show you what to do with it.

Make sure that the input signal from the signal generator is about 0.1 V. Set the gain of the amplifier to its lowest value (probably 1); the gain tells you how much the amplifier increases the size of the input signal. A System Alpha amplifier will increase the size of a signal by a maximum of about 10 times.

What is the size of the signal you get out of the amplifier?

Slowly increase the size of the input signal. What happens to the size of the output signal?

Increase the gain setting on the amplifier. Start with an input signal of 0.1 V again. Slowly increase the size of the input signal. What happens to the size of the output signal?

Keep a record of your results.

The gain setting on the System Alpha board is only a rough value. You can find the actual gain of the amplifier using this equation:

$$\text{gain} = \frac{\text{output signal size}}{\text{input signal size}}$$

Even if you haven't got a systems kit you can still do this activity. You will probably need to make your own amplifier. The type of amplifier you will want to make is called a non-inverting amplifier. It is quite simple to make one of these and a circuit diagram showing you how is given here:

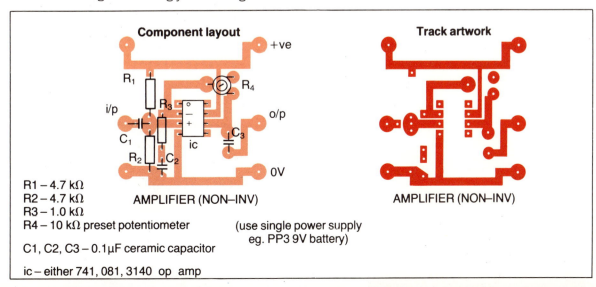

Component layout

+ve

R_1

i/p

R_3

R_4

o/p

C_1

ic

C_3

R_2

C_2

0V

AMPLIFIER (NON–INV)

Track artwork

AMPLIFIER (NON–INV)

R1 – 4.7 kΩ
R2 – 4.7 kΩ
R3 – 1.0 kΩ
R4 – 10 kΩ preset potentiometer

C1, C2, C3 – 0.1μF ceramic capacitor

ic – either 741, 081, 3140 op amp

(use single power supply eg. PP3 9V battery)

MIXING SIGNALS IN A STUDIO

In any modern radio station you will find a sound mixing desk. The controls on the desk are used to mix together different sounds like speech, music and special sound effects. These sounds may come from all sorts of different sources – a live rock band, a presenter or a pre-recorded tape. The signals can then be faded in and out, or superimposed on top of each other.

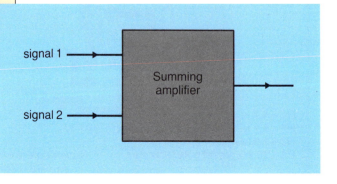

On the previous two pages of this book, you looked at amplifiers. Amplifiers can also be used to mix sound. If you have a systems kit, you will probably find it includes a summing amplifier. This device is called a summing amplifier because it adds together its input signals and amplifies (makes bigger) the result.

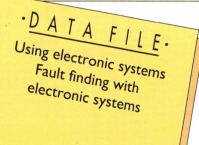

·DATA FILE·

Using electronic systems
Fault finding with electronic systems

If you haven't got a systems kit, you may need to make your own mixer. You will need a summing amplifier; the circuitry to make one is quite simple and is shown here:

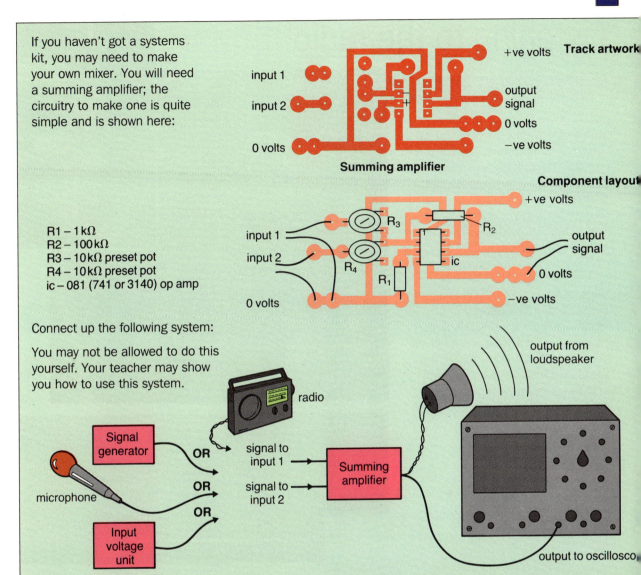

Track artwork

input 1

input 2

0 volts

+ve volts

output signal

0 volts

−ve volts

Summing amplifier

Component layout

R1 – 1 kΩ
R2 – 100 kΩ
R3 – 10 kΩ preset pot
R4 – 10 kΩ preset pot
ic – 081 (741 or 3140) op amp

input 1

input 2

0 volts

+ve volts

output signal

0 volts

−ve volts

Connect up the following system:

You may not be allowed to do this yourself. Your teacher may show you how to use this system.

Signal generator

microphone

Input voltage unit

radio

OR

OR

OR

signal to input 1

signal to input 2

Summing amplifier

output from loudspeaker

output to oscilloscope

Look at the output on the oscilloscope screen. Listen to the sound coming from the loudspeaker. Change the size of the signal at input 1. Watch what happens on the screen. Listen. Change the size of the signal at input 2. Watch what happens on the screen. Listen.

Connect the microphone to input 1 and talk. Watch what happens on the screen. Connect the radio output to input 2. Talk into the microphone and watch what happens on the screen.

You should see that, in each case, the two signals are superimposed on each other. Using the two level controls to change the size of the signals at the inputs, you can fade out the input signals or increase their strength. In this way you can add a 'voice-over' to music that is being played. You could also fade music in, over a script that you may be reading.

Suppose you are a DJ on a commercial radio station. A local supplier of sports equipment wants you to advertise their new range of sports bags. Write a script for a radio advert, choose some suitable music, and produce a 30 second advert. You may want to fade certain pieces of music in and out, perhaps to introduce and finish the advert.

Listening to the radio

SENDING RADIO SIGNALS

Radio waves are electromagnetic waves. Radio waves, which you can't see, are like light waves, which you can see. The difference between light waves and radio waves is the rate at which the waves vibrate.

How fast can you wave your hand backwards and forwards? How fast does a bee's wing vibrate?

Radio waves vibrate even faster at anything between 100000 vibrations per second to 100000000 vibrations per second. The number of vibrations per second is called the **frequency** and is measured in **Hertz** (Hz). Take a look at the dial or tuning panel on a modern radio. You will be able to see numbers representing frequency with the Hz symbol after them.

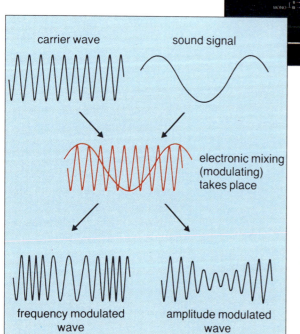

carrier wave

sound signal

electronic mixing (modulating) takes place

frequency modulated wave

amplitude modulated wave

Look at the radio again. There are two ways of transmitting radio signals: by *frequency modulation* and by *amplitude modulation*. You will often also see the letters FM, MW, and LW. These stand for Frequency Modulation, Medium Wave and Long Wave. Both MW and LW are examples of Amplitude Modulation or AM.

You cannot turn your voice directly into a radio wave and transmit it. Your voice must first be changed into an electronic signal and then added to a carrier wave. The carrier wave is a high frequency radio wave. The way in which your voice signal is mixed with the carrier wave decides whether a radio transmission is an AM or an FM transmission. This mixing process is called modulating.

Amplitude modulation is easier to achieve and show. You can produce an amplitude modulated signal for yourselves. You will need two signal generators, a summing amplifier and an oscilloscope.

Put this system together:

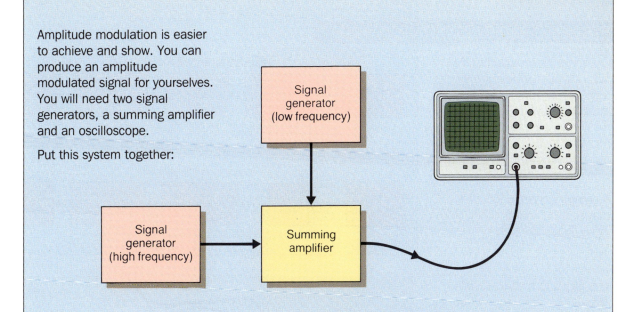

You should be able to see on the oscilloscope screen that the high frequency signal (the carrier wave) 'carries' the lower frequency signal. This combined signal can then be amplified and broadcast from a radio station's transmitting aerial.

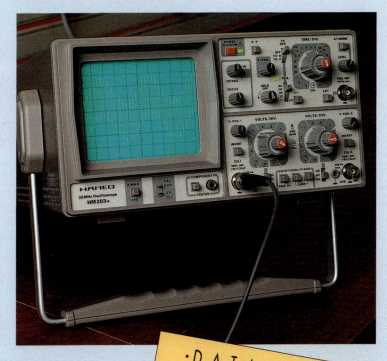

If you haven't got a systems kit, then you may need to make your own mixer. You will want a summing amplifier. You will find instructions on how to make one of these on page 75.

·D A T A F I L E·
Using electronic systems
Fault finding with
electronic systems

RECEIVING RADIO SIGNALS

Once you have transmitted a radio signal, the next problem is to receive it. A radio receiver must demodulate, or remove, the carrier wave to leave only the voice signal. The voice signal must then be amplified before going to a loudspeaker.

A complete system for transmitting and receiving radio waves looks like this:

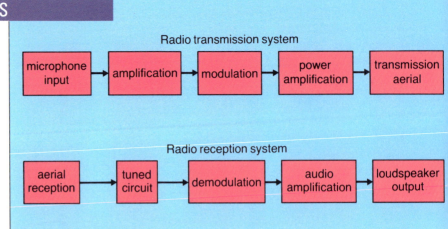

It is illegal to broadcast radio signals over a distance without a transmitting licence, so you will have to make do with making a radio receiver. If you have an electronic systems kit, this is quite straightforward. You can make an AM radio receiver with System Alpha like this:

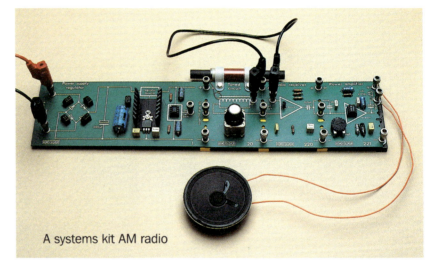

A systems kit AM radio

You might like to look at the signals that your radio is picking up. You might also like to look at the signals that parts of the receiver system are producing. Use an oscilloscope to investigate the signals produced at the different points in the system.

If you haven't got a systems kit, then you will need the following basic electronic circuitry to receive AM radio signals:

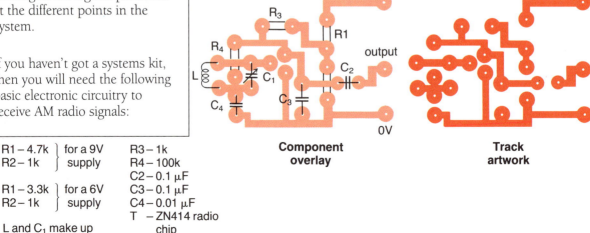

Component overlay

Track artwork

R1 – 4.7k ⎫ for a 9V	R3 – 1k
R2 – 1k ⎭ supply	R4 – 100k
	C2 – 0.1 μF
R1 – 3.3k ⎫ for a 6V	C3 – 0.1 μF
R2 – 1k ⎭ supply	C4 – 0.01 μF
	T – ZN414 radio
L and C_1 make up the tuned circuit	chip
tuned circuit is made from:–	50 turns of wire on a ferrite rod
	tuning capacitor (variable around 126 pF)

A new commercial radio station has begun operating in and around Cambridge. The station is called 'Cambridge Connect'. They have organised a competition, linked to a promotional drive for the station. The radio station would like *working designs* for new miniature radios to be given away. The winner will receive a free trip around the radio station and a punting trip on the river in Cambridge.

· D A T A   F I L E ·

Materials
Making a time plan
Using electronic systems
Fault finding with
electronic systems

There is a range of different activities in this section of the book. All of these activities are related to some aspect of computer control within the world of performance. Your teacher can give you a clearer picture of how the various parts of the section fit together. Some pages contain clearly set tasks, on other pages there are some fairly short activities, and on others there are opportunities for longer projects.

UNDER THE
Spotlight

We can mean many different things when we talk about control. So what is control? Sometimes we say that we have things under control. There are many activities that depend on having things under control. Gymnasts, for example, must have very good control of their bodies. Sometimes, however, things can get out of control. This usually means that, no matter what we do, we cannot prevent something from happening. You may have had an accident on a bike and fallen off because you lost control of it. Maybe you have had an argument and lost control of your temper.

If you look, you will find examples of control systems everywhere. Not all control systems are to do with human beings. There are also many examples of control systems in the natural world. Populations of animals, plants and insects tend to rise and fall depending on the availability of food, the weather, number of predators, and all sorts of other things. The natural world is controlled by all sorts of factors.

Your digestive system is another control system. If you are hungry, you will want to eat, and you will want to stop eating when you are full because your stomach will tell your brain that you should stop eating. Usually people don't eat too much — their eating does not get out of control.

In this chapter you will be looking at a special type of control — control using computers. Computers do many jobs that human beings would find boring and repetitive. They can be used on a production line in the manufacture of cars or televisions. They can be used to pack products and stack them in boxes. They can be used to analyse hundreds of medical and scientific samples at a time. In this chapter you will be looking at how computer control can play a part in putting on a performance.

lighting

Lighting is one of the most important parts of any production or performance. TV programmes, theatrical productions, films and concerts need to be lit to be seen. In each of these examples lighting can be used in a variety of different ways. A production or performance may use changing lighting effects. These changes in lighting will need to be controlled. Lighting is one aspect of production that can be easily controlled by computer.

The programmes that you watch on TV are often very carefully illuminated by many different types of lights. In a TV studio light levels are usually very high so that the TV cameras can take the best possible pictures. If you look at pictures of the inside of a TV studio, you will see lots and lots of different types of light.

Very strong lights are also used on a film set to enhance the effects of daylight and to remove unwanted shadows. They can also be used outside the studio to extend the working day.

In pop concerts lighting is often used to create special effects. Laser light shows may use beams of light to trace patterns in the air. Revolving mirrored glass globes may be used to produce speckled patterns over the musicians and audience. The lights may flash on and off to the sound of the music.

Lighting is also very important in the theatre. Lighting can be used to create special effects. It can be used to create the impression of a particular mood, atmosphere or time of the day. In the theatre, lighting is an essential part of any production. Lighting is not only used to illuminate the scenery but is also used to illuminate the performers.

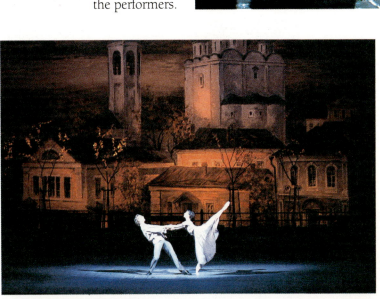

Nowadays, these sorts of lighting effects are often computer controlled. The lights are usually plugged into circuits which can be switched on an off by a computer. The computer can change light levels by dimming or brightening the lights. It can also change the colour of light by automatically moving different coloured filters in front of the lights. This can change the colour of a background or the mood of a scene.

switching On and off

Different schools have different computer systems. Different computers systems use different languages for control. Your school will have its own computer system. It might use any one of a number of control languages. These are some of the more commonly used languages: Contact, BBC Basic, Control Basic, Logicator, Control Logo, LEGO Lines, LEGO Logo, Controller, BITS, Control IT, and MCL.

IMPORTANT

The following pages contain several examples of computer programs for you to try. You will need to change these programs to suit your school's computer system and control language. Your teacher will tell you which control language your school uses and describe some of the more simple commands.

Your computer system will probably look something like that above.

Your computer can control equipment, like lights, through signal lines. Many of the control languages listed above show what is happening to the signal lines – whether they are off or on – by a display on the screen.

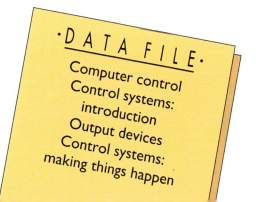

·DATA FILE·

Computer control
Control systems: introduction
Output devices
Control systems: making things happen

IN — OUT
7 6   5 4 3 2 1 0

? TURN ON [0 1 2]

Signal being sent out on lines 0, 1 and 2

No input signal on lines 6 or 7

Connect up this system:

You can use the computer to switch the lamp (or output device) in this circuit on and off. Often a simple command is all that is needed to switch on the output device:

`SWITCH ON 1`

Switching on more than one device in one go can also be simple:

`SWITCH ON 1 2 3`

Switching off the devices is as easy:

`SWITCH OFF 1 2 3`

Some control languages will let you switch off all the signal lines in one go, using a command such as `RESET` .

Try out some of these commands on your system. Try connecting different output devices to your computer interface.

You may want to switch on different devices at different times, so you may want the computer to wait before switching on each device. A `PAUSE` or `WAIT` command can be used to indicate how long you want the computer to wait. The time could be specified in seconds or in tenths of a second.

Here is an example of a program which uses this:

`SWITCH ON 1`

`WAIT 5`

`SWITCH ON 2`

`WAIT 3`

`SWITCH ON 3`

`WAIT 10`

Can you describe in words what this program will do?

procedures

You can't always just type a program like this in and expect it to work. You have to make the commands into a program, called a **procedure**. Procedures are simply bits of programming that perform specific functions. Each procedure must be given a separate name.

You may be able to generate a procedure using commands like `BUILD` or `TO`. Try typing this instruction into the computer:

`BUILD LIGHTON`

(You may find the screen display changes when you type in the command to build a procedure.)

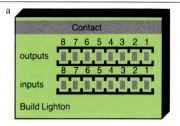

a

Contact
outputs 8 7 6 5 4 3 2 1
inputs 8 7 6 5 4 3 2 1
Build Lighton

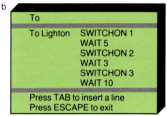

b

To
To Lighton   SWITCHON 1
WAIT 5
SWITCHON 2
WAIT 3
SWITCHON 3
WAIT 10
Press TAB to insert a line
Press ESCAPE to exit

Now type in the short program from the end of the previous page again. You will now find that every time you use the command `LIGHTON`, that short procedure will run.

Now try building a procedure to switch the lights off again. Call it `LIGHTOFF`. Start with `BUILD LIGHTOFF`. Now over to you!

You now need only two commands – `LIGHTON` and `LIGHTOFF` – to control the lights. Try making up procedures of your own to switch on different output devices – buzzers and motors for instance – in different orders. The procedures that you build may be used not only on their own, but also within other procedures. For instance, you might decide to build a procedure called `ACTION` which looks like this:

`BUILD ACTION`

`LIGHTON`

`LIGHTOFF`

Now, when you type `ACTION`, the two procedures which you have already written – `LIGHTON` and `LIGHTOFF` – will run one after the other.

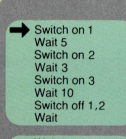

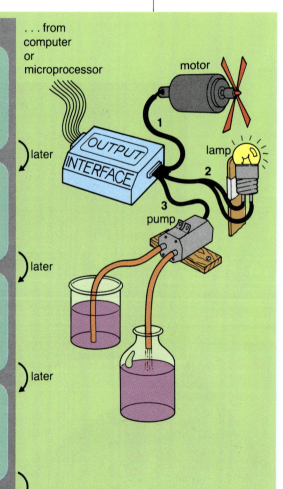

. . . from computer or microprocessor

motor

lamp

OUTPUT INTERFACE

1

2

3

pump

Switch on 1
Wait 5
Switch on 2
Wait 3
Switch on 3
Wait 10
Switch off 1,2
Wait

) later

Wait 5
Switch on 2
Wait 3
Switch on 3
Wait 10
Switch off 1,3
Wait 7
Switch off 2

) later

Wait 3
Switch on 2
Wait 3
Switch on 3
Wait 10
Switch off 1,3
Wait 7

) later

Switch on 2
Wait 3
Switch on 3
Wait 10
Switch off 1,3
Wait 7
Switch off 2

)

Wait 3
Switch on 3
Wait 10
Switch off 1,3
Wait 7
Switch off 2
End

repeating procedures

You may want something to happen more than once. This could mean that you want a procedure to run several times or even that you want it to keep repeating itself indefinitely. You can do this by including the name of a procedure within the procedure itself. A procedure of this type might look like this:

`BUILD LIGHTON`

`SWITCH ON 1`

`WAIT 5`

`SWITCH ON 2`

`WAIT 3`

`SWITCH OFF 1`

`WAIT 10`

`SWITCH OFF 2`

`WAIT 4`

`LIGHTON`

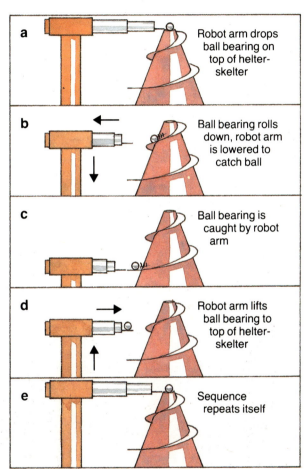

a Robot arm drops ball bearing on top of helter-skelter

b Ball bearing rolls down, robot arm is lowered to catch ball

c Ball bearing is caught by robot arm

d Robot arm lifts ball bearing to top of helter-skelter

e Sequence repeats itself

When this procedure has run the last thing it does is to call up `LIGHTON`. So the procedure starts itself up again when it reaches the end and it will go on repeating itself for ever if you don't unplug the computer or tell it to stop!

Another useful command you may want to use is `REPEAT` or `REPEAT UNTIL`. This can be used if you want your procedure to run more than once but you don't want it to go on running for ever. Repeating a procedure is like going around a loop several times. You could build a program which looked like one of these three examples:

`REPEAT 120`

(instructions)

`END`

`REPEAT UNTIL` (condition)

(instructions)

`END`

`REPEAT`

(instructions)

`UNTIL` (condition)

The `REPEAT` command can also be built into procedures and you can put procedures within `REPEAT` loops. Now try using this command in your own programs.

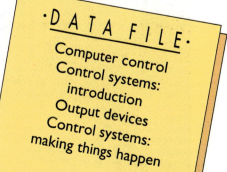

·DATA FILE·

Computer control
Control systems: introduction
Output devices
Control systems: making things happen

Writing a real control program

Believe it or not, you are now in a position to write a control program that would be useful to a real lighting designer in theatre or television studio!

You are responsible for the lighting for a play or a TV programme. Different scenes will need to be lit in different ways. Some scenes need a full bank of bright lights. Other scenes need only a small amount of light to create a dimly lit area or will need only a single spotlight for a small area of bright light. In some scenes coloured filters will be needed to produce special effects.

Imagine there are about five different scenes to be illuminated.

You will need to decide:
- which lights will be on in which scene,
- the timing between scenes,
- what colour effects you want,
- whether the performing area is ever going to be unlit,
- how long each light will be on for.

Write a control program that will automatically switch the lights on and off between the scenes.

Special lighting effects

Many exciting effects can be created using lighting. You may already have seen how lighting can be used to change the mood or atmosphere on a stage or set: blue lighting can be used to make a scene feel cooler; very faint blue light can actually produce a crisp white effect; red lighting can be used to 'warm-up' a scene. Some shades of red can even help to highlight skin colours. You might like to experiment with different coloured filters in front of a light.

Light can also be used to project shadows onto a set. This can be done by placing a mask inside a lamp housing, in the focal plane of the lamp lens system. The lens system is then used to focus the light over a particular area of the set.

Masks of this type are called a 'gobos' and are made out of metal. A metal gobo can get very hot inside a lamp housing. In theatrical lights gobos may even get so hot that they glow red. This is because the gobo is placed where the heat in the lamp is most concentrated.

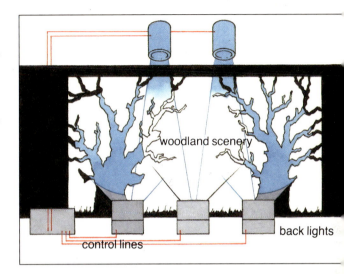

woodland scenery

back lights

control lines

All sorts of effects can be produced using gobos – star patterns, window frame shadows, woodland scenes – you can probably think of lots of other ways in which gobos could be used to produce special lighting effects. Try designing your own gobos to produce some interesting lighting patterns for a stage or set.

· D A T A   F I L E ·
Computer control
Control systems:
   introduction
Output devices
Control systems:
   making things happen
Materials

making light of control

In a real TV studio or theatre the lights change on certain cues. The lighting technician will be waiting for either audible or visual cues. He or she may also be listening for cues from the stage manager. The lighting technician could then flick a switch or press a button and switch on a certain arrangement of lamps or start a lighting sequence.

The lighting technician's script might look like this:

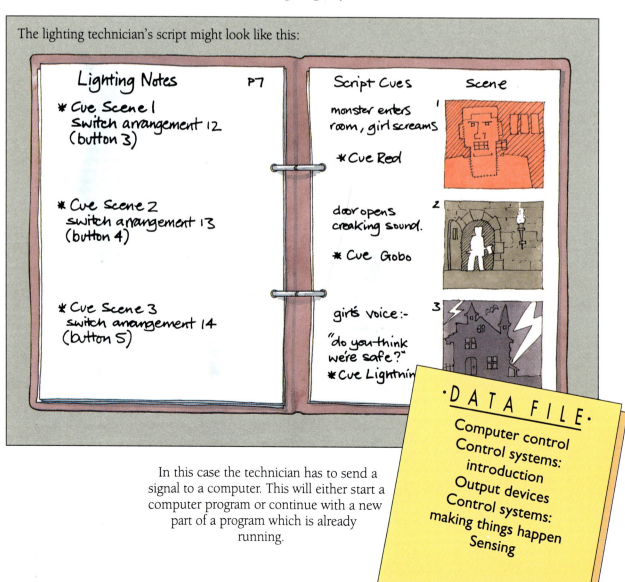

Lighting Notes   P7

* Cue Scene 1
  Switch arrangement 12
  (button 3)

* Cue Scene 2
  switch arrangement 13
  (button 4)

* Cue Scene 3
  switch arrangement 14
  (button 5)

Script Cues   Scene

monster enters   1
room, girl screams

* Cue Red

door opens   2
creaking sound.

* Cue Gobo

girl's voice:-   3

"do you think
we're safe?"
* Cue Lightnin

In this case the technician has to send a signal to a computer. This will either start a computer program or continue with a new part of a program which is already running.

· D A T A   F I L E ·
Computer control
Control systems:
introduction
Output devices
Control systems:
making things happen
Sensing

Many computer control languages have an input command; this could be something like `INPUT (number)` or `IN (number)` . The input command can be used to send signals into the computer. It is often used with another command, `IF` . The IF command is used to decide whether a certain condition exists or not. It could, for example, be used to check whether a button has been pressed, or whether an actress is ready or whether a set time has passed.

Here are some ways in which 'IF' could be used:

● IF the button is pressed then switch on lamps 2, 4 and 7,

● IF someone is standing on the pressure pad then switch on the spotlight.

This is the same as saying: if a condition exists do this, otherwise do something else.

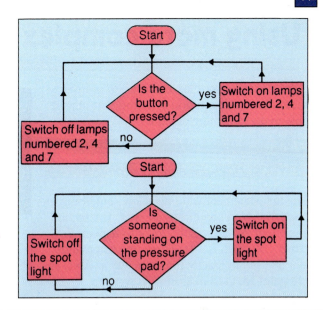

Let's start with a simple example.

Set up this system:

Suppose you want the lamp to come on when the button is pressed. You could write a procedure:

`BUILD SWITCH`

`IF INPUT(1) is ON: SWITCH ON 1`

`IF INPUT(1) is OFF: SWITCH OFF 1`

`SWITCH`

Now you can simply run `SWITCH` .

First, write a control program to switch the light on when the button is pressed. Now alter your program to switch the light on when the button is not pressed.

Connect other lamps, buzzers or motors into your system as output devices. Try switching two or three devices on together. Try switching them on or off in sequence. Try introducing delays between switching on different devices.

Now look at the lighting technician's script on the previous page. Write a control program that will operate the lights as required.

Remember:

● you will have three different inputs to monitor.

● you may only need to have some of the lights at any particular time.

● you will have to switch off lights you don't want.

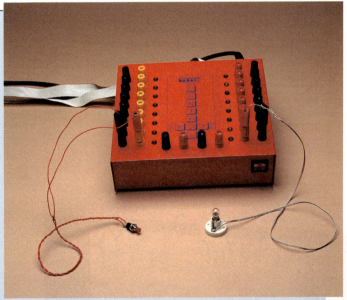

air pump

windscreen washer pump

solenoid-operated valve

Using more complex computer commands

Some control languages allow you to monitor more than one condition at a time. You may be able to do something like this:

IF INPUT (1) AND INPUT (2) are ON : SWITCH OFF 2 4 5

IF INPUT (1) OR INPUT (2) are OFF : SWITCH OFF 2 4 5

The first statement will only allow something to happen if both conditions are met. The second statement will allow something to happen if either condition is met. These two commands can be used in exactly the same way as AND and OR gates are used in electronic systems.

These are quite complicated statements. If you have a suitable control language you may be able to include statements like these in your control program. If you can, try out some of these ideas within the context of television or theatre lighting.

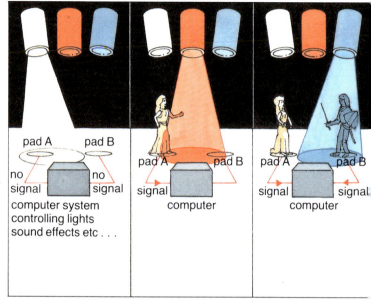

pad A   pad B

no signal   no signal

computer system controlling lights sound effects etc . . .

pad A   pad B

signal

computer

pad A   pad B

signal   signal

computer

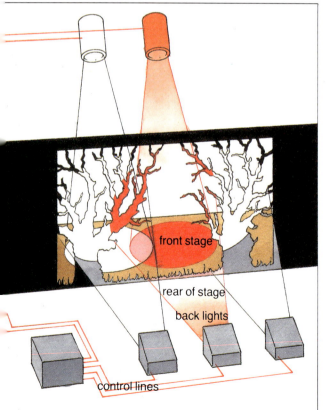

front stage

rear of stage

back lights

control lines

You might also be able to use more complex commands, that might be available with your control language, to:

● **alter the power level of an output**
  This would allow you to alter the light intensity of a lamp gradually. So you would be able to fade out or increase the brightness of lighting. You would also be able to gradually change the speed of a motor. Some control languages will allow you to do this using a command such as SETPOWER .

● **check the level of an input signal**
  Normally when you check for a signal, it is either present or not. Sometimes, if a signal is present you may want to know how strong it is. For example, if part of a control program depended on how strong the daylight was then you would want an accurate measure of light level. Alternatively, you might want to switch different devices on depending what temperature it was. Some control languages will allow you to do this using a command such as ADVAL .

Using more complex computer outputs

You are not limited just to switching lamps, buzzers and motors on or off. You could also use a computer to control water or air pumps, fans or pneumatic solenoids.

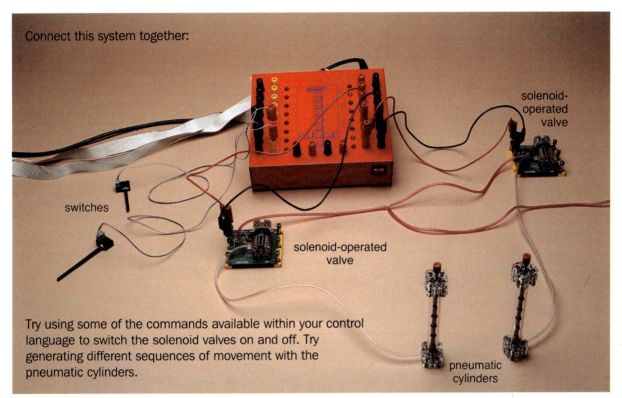

Connect this system together:

solenoid-operated valve

switches

solenoid-operated valve

pneumatic cylinders

Try using some of the commands available within your control language to switch the solenoid valves on and off. Try generating different sequences of movement with the pneumatic cylinders.

Here are some other ideas for getting movement with pneumatics.

Computer control system

The treasure awaits in here!

press me

press me

pressure pad

·DATA FILE·

Computer control
Control systems:
introduction
Output devices
Control systems:
making things happen
Sensing

LIVERPOOL HOPE UNIVERSITY COLLEGE

Here are some activities for you to try. Can you design these scenes for a TV or theatre production?

Charlie Bucket stared around the gigantic room in which he know found himself. The place was like a witch's kitchen! All about him black metal pots were boiling and bubbling on huge stoves, kettles were hissing and pans were sizzling and strange iron machines were clanking and spluttering, there were pipes running all over the ceiling and walls, and the whole place was filled with smoke and steam and delicious rich smells.

Aleph and Goma had moved carefully up the side of the mountain. Through a gap in the rocks they noticed a metallic door. Next to the door, glowing in the failing light was a large button. They pressed the button and the door slid open with a smooth swishing sound. Inside the cavernous room was a frenzy of activity. Pistons were moving here and there. Hissing sounds came from every corner. Lights were flashing on and off in complicated patterns and colours. They moved into the full glare of the lights in the centre of the room. There in front of them was a conveyor belt. Dropping, four at a time, into large boxes at the end of the belt were the products of all this commotion. "So this is where they're made!"

The lightning crackled outside over the bleak, dark moor. Flashes of light shattered the eerie darkness of this damp and dreary castle. The light streaked across the floor casting a shadow of the leaded window panes. Momentarily there was a glimpse of a strong wooden box in the corner of the room. The lightning flashed again, with an ear-splitting crack of thunder, revealing the box lid slowly opening and a jewelled hand feeling, fingering its way into the night world. The next flash of light came not from the storm raging outside but from the eyes of the creature that, even now, was emerging from its slumbers.

A Japanese TV company is planning to make a science fiction series based on 'Dr Who'. They plan to call this 'Doctor Ho'. It is due to be made next year. The company are having a competition to design new characters based on the Cybermen from previous series. The new series is to be called *The Return of the Cybertrons*.

You can use all the other sections in this book to get ideas.

·DATA FILE·

Computer control
Control systems: introduction
Output devices
Control systems: making things happen
Sensing
Materials

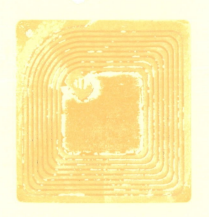